THE TURNING POINTS of
ENVIRONMENTAL HISTORY

The Turning Points *of* Environmental History

ಌ *Edited by* Frank Uekoetter

University of Pittsburgh Press

Published in cooperation with the
Rachel
 Carson
Center

Published by the University of Pittsburgh Press, Pittsburgh, Pa., 15260
Copyright © 2010, University of Pittsburgh Press
All rights reserved
Manufactured in the United States of America
Printed on acid-free paper

10 9 8 7 6 5 4 3 2 1

Library of Congress Cataloging-in-Publication Data
The turning points of environmental history / edited by Frank Uekoetter.
    p. cm.
  Includes bibliographical references and index.
  ISBN 978-0-8229-6118-5 (pbk. : alk. paper)
  1. Environmental sciences—History. 2. Human ecology—History. 3. Environmentalism—History. 4. Environmental degradation—History. 5. Nature—Effect of human beings on. I. Uekötter, Frank, 1970–
  GE50.T87 2010
  333.7209—dc22   2010031724

For Richard H. Grove

Contents

	Acknowledgments	ix
1.	Thinking Big: The Broad Outlines of a Burgeoning Field **FRANK UEKOETTER**	1
2.	The First Hundred Thousand Years **J. R. MCNEILL**	13
3.	Agriculture **DEBORAH FITZGERALD**	29
4.	Forest History **BERND-STEFAN GREWE**	44
5.	The Nation-State **THOMAS LEKAN**	55
6.	Urban Environmental History **JOEL A. TARR**	72
7.	The "1950s Syndrome" and the Transition from a Slow-Going to a Rapid Loss of Global Sustainability **CHRISTIAN PFISTER**	90
8.	Modern Environmentalism **JENS IVO ENGELS**	119
9.	The Knowledge Society **FRANK UEKOETTER**	132
10.	Desertification **ALON TAL**	146
	Notes	163
	List of Contributors	201
	Index	203

Acknowledgments

This book grew out of a workshop held in June 2005 at the Bielefeld University Center for Interdisciplinary Research. The editor would like to thank Christof Mauch and Joachim Radkau for organizing this event, as well as the Center for Interdisciplinary Research and the German Historical Institute in Washington, D.C., for financing it. Thanks are also due to all the presenters and participants who made the Bielefeld workshop such a stimulating experience that called for a follow-up. Special thanks to the authors, who dared to turn provocative "thought pieces" into full-length chapters. Two authors bravely joined the ongoing project. Two anonymous reviewers gave critical input, thus helping to refine the ultimate product. Karen Oslund was a great help in improving the prose of non-native English speakers. Elisabeth Stevens took care of the index.

In publishing this book, the University of Pittsburgh Press proved to be a reliable partner. As it happens, the conclusion of this project coincides with the start of the Rachel Carson Center in Munich, Germany, a new institute for advanced study that will investigate many of the issues discussed herein in great depth. Environmental history has clearly grown beyond its childhood days, but the field is learning that adolescence has its problems as well. If this book helps to tackle some of them, it shall have fulfilled its purpose.

THE TURNING POINTS of
ENVIRONMENTAL HISTORY

1

Thinking Big
The Broad Outlines of a Burgeoning Field

FRANK UEKOETTER

❦ It has become conventional wisdom among scholars that environmental history has grown up. From a marginalized field caught between counterculture activism and professional rigor, it has developed into an established part of the scholarly community that no self-respecting history department would ignore. Environmental history meetings routinely attract audiences in the hundreds, the number of books and journal articles on the topic has expanded enormously, and the field has matured in methodological terms as well. Whereas declensionist narratives were still a powerful current only two decades ago, declensionism now features as a subject entry in Carolyn Merchant's *Columbia Guide to American Environmental History*.[1] Environmental historians know well that growth usually comes at a cost, however. The field has certainly benefitted from this boom, but the growth has also changed its general character. Most crucially, studies have become more specialized in recent years. While the first generation of environmental history books often covered centuries and multiple countries, or even continents, recent studies are more limited in chronological and geographic terms.

This development is certainly not peculiar. Specialization is a frequent trend in booming fields, and arguably an inevitable one, and it does have its merits. Few scholars would doubt that a growing specialization of environmental history research has helped to boost a diversity of themes and methodologies, which has made the discipline richer than ever. But at the same time this boom is making the big trends of environmental history increasingly obscure. For the first generation of scholars, environmental history was not just a set of case studies but a fundamental challenge to our understanding of history. This challenge is still there, although somewhat hidden under a growing pile of specialized studies. Environmental history is more than a cluster of interesting fields: it points to the need for more comprehensive thinking about the past that includes animals and plants, the land, the sea, and the atmosphere, and to the wide range of ideas and practices that link these entities with human societies. It is time for scholars to survey the field in a broad way to widen our view from case studies to the key trends that now define the field. And that is what this book intends to do.

Turning Points

For environmental historians seeking a broader understanding of history, turning points are a great way to start. After all, the implications of turning points are enormous: they define time frames and chronologies, they highlight certain trends at the expense of others, they provide structure and focus—in short, turning points provide a backbone to narratives that no scholarly study of history can do without. Discussions over periodizations have been an enduring concern among students of history, a perfect way to stimulate conversation between subdisciplines. For all the diversity of scholarship today, most historians still agree that turning points are important. In fact, when it comes to environmental issues, looking at turning points is far more than an academic endeavor. The issue goes right to the heart of environmental thinking: since the nineteenth century, notions of "decline" and "renewal," of a "fall from grace," and of a "turnaround from the brink" have permeated the environmental discourse and continue to resonate in modern environmentalism. Countless environmental initiatives have been touted as "a watershed" or even a "last chance," invoking a notion of prospective turning points that has spurred many laws and other measures. Even the dystopias and horror scenarios so dominant in environmental discourses mirror thinking in grand chronological schemes. It is difficult to talk about environmental issues without talking about turning points.

Returning to the big questions hanging over environmental history begs us to reconsider some of the classic studies in the field. Specifically, it deserves attention that some of the early landmark publications were essentially reflections on turning points. A classic example is Lynn White's famous essay on "the historical roots of our ecological crisis," which put the blame for environmental destruction on the rise of Christianity, with the honorable exception of St. Francis.[2] A second example is Rolf Peter Sieferle's attempt to structure world history with a view to its energy base, which had a solar energy regime dominating since the

Neolithic Revolution, to be replaced by a new one based on fossil fuels in the wake of the Industrial Revolution.[3] These are two of the better-known arguments for turning points; other authors saw the crucial shifts in antiquity or in eighteenth-century Enlightenment thinking.[4] But even where chronologies differ, it is worth noting that these studies adhere to the same concept of turning points. For the first generation of environmental historians, the turning points of environmental history were essentially shifts from an "ecologically benign" (that is, "good") life to a "destructive" (that is, "environmentally unsustainable") one.

It goes without saying that these early histories played an important role in stimulating further research on environmental issues. It would be arrogant indeed to look down from a twenty-first-century perspective on these first attempts to define the field and chastise their naïveté. Historiographic revolutions are inevitably bound to start with daring assumptions and speculations. It is equally clear, however, that these grand narratives have fared badly. Many critics have focused on important details, such as the disputed "wood scarcity" that figured prominently in Sieferle's interpretation.[5] But beyond these specifics, two general problems stand out in hindsight. One is the issue of monocausalism. Early narratives routinely sought to identify a single cause for environmental decline: energy resources, environmental thinking, and so on. The search for the single factor that put the environment on a downward slope was bound to prove elusive, and corresponding attempts were regularly defied by the complexity of history.

The second general problem was the Manichean structure of the arguments: they made a sharp distinction between "before" and "after," with environmental stability being given before the turning point and monotonous decline thereafter. In short, the turning point was usually a fateful "point of no return," a secular watershed that had the course of history flowing in only one direction henceforth. Therefore it was tempting to dismiss these periodizations as overly simplistic, and it seems that this is what the field has been doing ever since. Most recent studies no longer discuss these general turning points; rather, they abandon the topics altogether. In doing so, environmental history may have lost more than it has realized.

Of course, one should be mindful of the shortcomings of these early proposals when reopening the debate over turning points. In the twenty-first century the periodizations of environmental history should be more complex, more nuanced, and leave more room for opposite and multidirectional trends. But in what way? The search for overarching periodizations is exceedingly difficult, and turning points can be identified more easily in specific, broadly conceived fields of research like urban history or agricultural history. The contributors to this volume were asked to respond to two general questions: (1) Where do you see the key turning points in your respective field of scholarship? (2) What are the reasons for your choice? Needless to say, this forced choice gave headaches to the contributors, and some of this uneasiness can be felt in the chapters. John McNeill put it most eloquently when he notes in his chapter that he rushes "where prudent angels fear to tread, to the realm of long-term global-scale history." If this book were a medication, it would certainly need to include a legal disclaimer: "Warning! This book contains sweeping generalizations and significant omissions!"

In defense of the overall enterprise, the authors did approach the topic cautiously. The issue was first discussed during a workshop at the Bielefeld University Center for Interdisciplinary Research in June 2005. In light of the topic's trickiness, participants were asked to provide short "thought pieces" of some three pages, which they were to present and defend during the conference. After lively and stimulating discussions, many participants felt that the debate should be continued, and that is how the idea for this book was born. The majority of the following chapters are essentially revised and expanded versions of these informal thought pieces; some chapters were added to broaden the focus of the volume. The contributions are open to debate at many points and from numerous angles, and readers will appreciate the authors' courage to draw clear lines when it would have been easier to mutter about "complexity" and "difficulties." The contributors were encouraged to provide minimal annotation—no plethora of chapter notes could possibly mirror the richness of the fields in question, although some authors have tried their best nonetheless.

An endeavor of this kind inevitably hinges on a balanced selection of topics, which was difficult to achieve. At the risk of stating the obvious, the list of themes included herein is by no means comprehensive. In fact, any claim at comprehensiveness would be defeated by the absence of a universally agreed-upon definition of what "environmental history" in fact is. As McNeill has remarked elsewhere, the limits of environmental history are, as disciplinary boundaries go, "especially fuzzy and porous."[6] He has identified three main fields of environmental history inquiry, and this book made a point of looking at all three: the material (the forests); the cultural/intellectual (the knowledge society); and the political (the nation-state).[7] The contributions also cover such topics as agriculture and urban environmental history, which include all three fields. Emphasizing the costs of energy in Western societies after 1945, Christian Pfister's piece starts on the material level but adds a discussion of cultural and political implications. His phrase "1950s syndrome" has become a household term among environmental historians of Central Europe and probably deserves more attention in the English-speaking world. This book includes a chapter on the environmental revolution—a self-proclaimed turning point that, as a side effect, opened the historiographic niche for the discipline of environmental history. A chapter on the fight against desertification provides a much-needed antidote to a Western bias that all too easily creeps into discussions of world history. McNeill's chapter puts it all into an even bigger perspective with his grand discussion of "the first hundred thousand years."

There is no sense in dwelling on the omissions, but the absence of a chapter on the Industrial Revolution deserves a few words of explanation. There can be no doubt about the transformative potential of modern industrial capitalism, but three arguments ruled against such a chapter. First, the concept of the Industrial Revolution has recently come under fire from economic historians who insist that growth rates were lower than presumed for a long time; some have abandoned the concept altogether. Furthermore, industrialization was a decidedly regional phenomenon, as Sidney Pollard has argued, and the differences between regions pose enormous problems for a project that seeks to move beyond specific localities.[8] Finally, the Industrial Revolution was not a causative factor in its own right but merely the product of a complex confluence of trends. Research on the environmental impact of industrial companies has long shown that their behavior was not so much an outgrowth of a quintessential industrial point of view (that is, the desire to make money) as a reaction to certain institutional settings: to laws and state agencies as well as public pressure that encouraged or discouraged certain forms of behavior. With that, it seems more worthwhile to focus specifically on the agents that molded the behavior of industry, rather than to conflate them in a single chapter treating a bewildering mixture of factors.

Toward a Comprehensive Chronology

It would be a missed opportunity if we did not try to bring these chapters into a dialogue with each other, to identify common overlapping themes. One thing that emerges is that turning points in environmental history are not necessarily momentous events. Surprisingly, that does not mean that politics was generally unimportant from an environmental history perspective. This collection of essays shows that environmental historians have overcome the early antipolitical bias that Thomas Lekan mentions in his chapter. Instead, in various chapters the book gives credit to the Napoleonic Wars (Bernd-Stefan Grewe), the two world wars (Deborah Fitzgerald), and even single events like the Sandoz Fire of 1986

(Lekan). But eventually these events take a backseat, being important primarily as accelerators of ongoing developments or indicators of more general trends. Most turning points discussed herein are not points in a literal sense, as they extend over a certain period of time. Typically, the turning points of environmental history are periods of accelerated and correlated change in different areas.

Surprisingly, the recent boom of research on natural disaster did not leave its mark in this volume. Alon Tal's chapter is the only one in which disasters figure prominently. Tal details the creation of the U.S. Soil Conservation Service in the wake of the Dust Bowl of the 1930s, and he goes on to describe the Sahel famine of 1968 that defined desertification as an "African" issue. The other contributions pay little attention to catastrophic events; in fact, Joel Tarr's discussion of urban environmental history emphatically denies that disasters had any impact in the long run, although post-Katrina New Orleans may make for an exception. The contrast deserves some closer scrutiny. Should we take this as a sign that natural disasters are really less important in the long run than the current wave of scholarly interest suggests? Or is the underlying factor the geographic focus of this volume? Maybe frontier regions and Third World countries possess a vulnerability to disasters that makes them more prone to their impact than, say, large urban settlements?[9] Perhaps the general character of the field of environmental history plays a role here, too: with natural processes usually proceeding at a rather leisurely pace, long-term processes probably gain prominence in environmental history narratives. With that, the relative marginality of disasters writ large would be merely one facet of a general immunity toward sharp turning points: *natura non fecit saltus*, meaning "nature does not jump." It would be unwise to suggest a definitive answer at this point, but the scarcity of disasters as turning points in this book points to the need for further reflection.

The porous character of turning points in environmental history is especially strong when it comes to the early modern era. Of course, the Columbian Exchange that figures prominently in McNeill's synthesis makes for an exception given the disastrous epidemics that depopulated large parts of the New World and the introduction of new megafauna, but that only holds true for one area and one moment in time. After all, Alfred Crosby made a point of stressing the divergent speeds of the transformations in America and the rest of the world. Whereas the new continent changed dramatically, transfer processes in the Old World were much slower and geographically diverse.[10] McNeill's chapter acknowledges these divergent chronologies, as he presents the year 1492 merely as the beginning of a biological exchange that continues through present day. The same holds true for McNeill's most recent turning point: the shift to fossil fuels. While traditional hagiographies of industry present the introduction of coal as a revolutionary step, McNeill says it was really a prolonged process. It took several generations to get from wood-based economies to the dependency on coal, oil, and gas that figures so prominently in current political debates. At the end of many small steps, Western societies finally found themselves locked in a reliance on fossil fuels that nobody had envisioned, let alone desired.

The fuzziness of turning points in the early modern era is also evident in the history of forestry. For several centuries early modern rulers enacted forest ordinances, but these have only very slowly transformed forest use over time. The same holds true for the agricultural revolution in Great Britain, which stretched roughly between 1500 and 1750. It seems that turning points during the early modern period were to a much greater extent defined by the specific context of a certain country. Grewe, in his chapter, mentions the example of Colbert's nationalization of forestry in 1669 and the expropriation of church and nobility during the French Revolution, two defining events of forest history in France without much relevance beyond the country's borders. Likewise, the British agricultural revolution moved only slowly beyond the confines of the island, and the same holds true for the plantation economies in the Caribbean. It was not until the nineteenth century that turning points developed more in sync with each other, a trend easily attributed to the growing international exchange that led to what some have called the first era of globalization around 1900. In a way the rapid worldwide spread of modern environmentalism around 1970 was the endpoint of a global synchronization of turning points in environmental history.

Does the environmental history of the countryside ask for a periodization that differs from that of urban environmental history? My general impression is that the answer may be a cautious yes. Although the late nineteenth century emerges as a crucial time of change in the urban environment, the turning points in agriculture were both earlier and later. In fact, Deborah Fitzgerald's chronology for agriculture in this volume looks remarkably similar to Grewe's time line of forest history. In both cases crucial turning points occurred in the first half of the nineteenth century and in the second half of the twentieth, although the reasoning is by no means congruent. While Fitzgerald and Grewe both emphasize the impact of railroad construction, Tarr sees the streetcar as more important for urban environmental history. His case for the "networked city" as a key turning point in the environmental history of cities seems well argued, and yet it is striking how little urban events spread into the countryside, or into other chronologies for that matter. Thirsty cities may drain certain places of precious water, and dump their wastes on others, but the impact of these practices was spatially limited. Of course, that was little consolation for people in places like Owen's Valley, but the environmental implications of the modern city seemingly remained strongly urban in their reach until far into the twentieth century. Those who have read William Cronon's *Nature's Metropolis* may be tempted to object, but Chicago was probably less than typical in that respect.[11]

The turning points of environmental history were not necessarily characterized by direct changes in humankind's impact on the natural environment. Several contributions in this book stress developments that primarily changed the *structure* of environmental debates, rather than their environmental implications. In Lekan's discussion of the nation-state, it is the rise and fall of high modernism that defined the turning points. The nation-state emerged as the prime political actor of environmental policy during the nineteenth century and then gradually lost its grip after World War II. In other words, the key event was a penchant for centralized planning and expertise that was aloof from local concerns, rather than a specific impact on the environment. Likewise, I argue in my chapter on the modern knowledge society that although the direct implications of knowledge are too diverse to allow a summary assessment, the change in discursive patterns was uniform. Since the mid-nineteenth century, it has become increasingly difficult to participate in environmental debates without resorting to academic expertise. In Jens Ivo Engels's presentation of West German environmentalism, the much-touted "ecological turn" looks impressive only as a transformation of prevailing political styles.

With surprising stringency, the mid-nineteenth century emerges as a general watershed of environmental history, figuring prominently in the chapters by Fitzgerald, Grewe, Lekan, and my own. The link here to industrialization is weaker than one might guess. Fitzgerald stresses the application of science and technology to agricultural production as an important shift around 1850, but these were only modest beginnings in a process of industrialization of agricultural production that did not gather steam until the mid-twentieth century. In forest history a direct link is counterintuitive, as wood was not the favored fuel of the industrial era (though it continued to be an important building material). The rise of high modernism had much to do with improved communication links and richer state coffers, but the reforms of state administrations that inaugurated what the historian Charles Maier has called the age of territoriality had roots that were far more diverse.[12] Finally, the rise of academic expertise was certainly related to industrial development, and yet it was far more than its direct offspring. In short, the rise of industrial might looms through all four turning points, and yet it was only one part of a much broader story. As a force of environmental history, industrialization was a far more complex phenomenon than the first generation of researchers would have guessed.

The Twentieth Century: Environmental Perspectives

After a cluster of turning points in the nineteenth century, the first half of the twentieth century looks notably scarce of turning points. Tal points to the Dust Bowl and to the institutionalization of soil conservation expertise, while Tarr stresses the start of suburbanization in the 1920s. However, suburbanization did not gather steam until after World War II, making the scarcity of turning points in the first half of the twentieth century even more clear. This finding is all the more surprising because the early decades of the twentieth century were by all means eventful in political terms. Two world wars, the Great Depression, the collapse of many European democracies, and the unrest in the colonial world may serve to demonstrate this point. So how does one correlate a turbulent political history with an environmental history largely devoid of turning points?

A plausible argument might see the interwar years as something of a hiatus for turning points. With the Depression and the repetitive demands of war economies, the shift to consumer societies with severe environmental implications was postponed for several decades. A similar argument may be made for environmental policy. With the attention of policymakers being consumed by other topics, the interwar years lacked the long-term stability that usually preconditions environmental policy debates. The latter correlation, however, is by no means cogent. The New Deal was obviously a response to the Great Depression, and yet it helped to reinvigorate the conservation drive of the Progressive Era. It was instrumental for, among other things, the creation of soil conservation institutions with global reach. In Nazi Germany conservationists were jubilant after the new leaders pushed through a national nature protection law in 1935 that bore the hallmarks of an authoritarian state. But then, Nazi Germany was the only European country with a boom of conservation work in the 1930s.[13]

In general, environmental issues played more of a marginal role during the interwar years, and conservation movements were, with some exceptions, at bay. Environmental history followed pre-1914 trends in a more or less halfhearted way. The striking thing about the twentieth century as discussed here is the multitude of turning points *after* World War II and their scarcity *before* 1945. But how does one correlate this finding with McNeill's well-known argument in his *Something New under the Sun* that "the twentieth century was unusual for the intensity of change and the centrality of human effort in provoking it"?[14] One should probably not take McNeill's argument too literally, as his book freely refers to events in earlier centuries without much ado. A more worthwhile target for criticism may be the notion of a "long nineteenth century" so popular among European historians, which starts with the French Revolution and ends in 1914, to be followed by a "short twentieth century" that ended with the collapse of communism in Eastern Europe around 1990. This periodization obviously has its roots in political history, and it seems that it is of little use when it comes to environmental history. Instead, environmental historians may find it more useful to speak of a "long twentieth century" that started somewhere around 1850, with changes in agricultural production, new transportation networks, a growing importance of national institutions, and the environmental repercussions of urban growth. To be sure, the idea of a "long twentieth century" is not completely new. In a recent article by Will Steffen, Paul Crutzen, and John McNeill, the coauthors proposed to speak of an "Anthropocene"—a new chapter in the earth's geophysical history—because sometime around 1800 or 1850, humans were starting to overwhelm the great forces of nature. However, their definition hinges on a single indicator—the concentration of carbon dioxide in the global atmosphere—and it should be obvious that this choice was inspired by the current debate over global warming.[15] One might read this book as a call to put the debate on the Anthropocene on a much broader base.

Another striking feature is the bewildering mix of turning points in the years after World War II. Nearly all the chapters argue that crucial events took place during the second half of the twentieth century, but in a somewhat cacophonic way. Fitzgerald's discussion of agriculture and Pfister's argument for the "1950s syndrome" both emphasize a crucial shift toward more resource-intensive, exploitative practices. For Fitzgerald postwar agriculture saw a new intensity in the use of scientific and technological expertise, a boost in productivity, and a near-totality in the industrialization of agricultural production, which together marked a new stage in the environmental history of agriculture. Pfister sees the human ecological footprint expanding enormously since the 1950s, with cheap energy prices and new consumption patterns paving the way for unprecedentedly wasteful societies.

But when one moves from material consumption to politics, the picture becomes more confusing. A conventional narrative sees a conversion to the environmental cause taking place in Western societies around 1970, but that interpretation fares poorly in Engels's interpretation, which depicts the famous "ecological turn" as a rather limited event. In fact, two chapters in this book depict the environmental revolution as a rather unlikely event. Lekan notes that the age of high modernism was coming to a close around 1970, and the forces of economic globalization were gradually eroding the regulatory power of the nation-state. In my later chapter, the modern knowledge society reached the apogee of its power in the postwar years, with the trust in science-based expertise culminating in atomic-age fantasies.

So did modern environmentalism produce little more than—in Fernand Braudel's famous formulation—"surface disturbances, crests of foam that the tides of history carry on their strong backs"?[16] More specifically, were the truly crucial trends of the postwar years the transformations in material production, with the "age of ecology" being a merely ephemeral phenomenon by way of comparison? Grewe sees an impact of modern environmentalism on forest management, and Tarr envisions a new era of urban environmentalism beginning around that time as well. But those accomplishments clearly shrink in scale when one moves beyond the confines of Western industrialized countries. It is difficult to read Tal's narrative of the agonizingly slow progress in the fight against desertification without a feeling of remorse. If desertification is indeed the environmental problem affecting the largest number of people around the year 2000, as the 2005 Millennium Ecosystem Assessment contends, then why was it so difficult for African nations to put the topic on the agenda of the 1992 UN Conference on Environment and Development (UNCED) in Rio? In my reflections on the future of the knowledge society, I speculate that with the dependence on expert advice, it has become so difficult for environmentalists to separate important from unimportant issues that the "great environmental awakening" may one day look more like a "great deception." Perhaps that day has already arrived, at least on the issue of desertification.

Tal's essay ends on a cautiously optimistic note, as the fight against desertification has been gathering momentum in recent years. His discussion shows that the future of arid and semiarid lands may not lie solely in grand political schemes, but also in small changes with huge implications. The technology of drip irrigation emerges as one of the hidden heroes of environmental history in Tal's narrative, a force that was much more important to the fight against desertification than the wave of environmental activism in the 1970s. His chapter shows that environmental history, as a profession, has come a long way from its 1970s roots. From an emphatic embrace of environmental values and environmental activism, researchers have moved to a more balanced and even skeptical perspective on environmentalism, seeking to include accomplishments as well as failures in their narrative. We rarely hear statements from environmental historians nowadays that it is a "right of historians to be advocates and moral critics."[17]

But in the process we have certainly not become apolitical or devoid of political relevance. As it happens, this book was going to press just as environmentalists all over the world were stressing the importance of the climate summit in Copenhagen in December 2009, hoping that it would mark a turning point for global climate policy—the "last exit" from our fatal addiction to fossil fuels. This volume certainly offers an ambiguous impression in this respect: in the history of humans and the natural world, turning points were rarely political events. At the same time, however, humans have never been so tightly connected globally and so well informed about their ecological footprint. Turning points were usually unexpected and often came from a confluence of initiatives and efforts that even the wisest blueprint (or global climate deal, for that matter) could barely envision. We do not know when the next turning point of environmental history will come or what it will mean. But at the very least, we can be sure that the debate over turning points, politically as well as historically, will not end with this book. In fact, as we have tried to show, it has barely begun.

2

The First Hundred Thousand Years

J. R. McNEILL

◈ I rush in where prudent angels fear to tread, to the realm of long-term global-scale history. As a rule, historians leave this treacherous terrain to others, to historical sociologists in particular. Historians have their reasons for this caution, preferring the surer ground of smaller-scale history that can be supported by written documentation. One way to reduce the impracticalities of long-term global-scale history is to privilege one variety of human experience. I confine my scope here to environmental history, and to a few turning points within it. Of all the possible candidates for the title of "turning point" in the long history of human habitation of the biosphere, I maintain that seven stand out. They are, in chronological order, (1) the harnessing of fire; (2) the emergence of spoken language; (3) the exodus from Africa and occupation of other lands; (4) the domestication of plants and animals; (5) the emergence of cities; (6) the Columbian Exchange; and (7) the development of fossil fuels.

For most of the human career, we lived in small bands and roamed large territories. It is difficult to know a lot about these millennia, but much of what little we know has to do with human involvement with the environment. It was surprisingly eventful. First, at some point our ancestors, whether human or hominid, learned how to harness fire and then to make it. This was one of the great turning points of human history, not merely environmental history, although we do not know when or where it happened. Informed guesses suggest this occurred maybe half a million years ago, which is before the appearance of *Homo sapiens*.[1] It almost certainly happened somewhere in Africa, where most hominids were

13

and where *Homo sapiens* first emerged. The use of fire allowed our ancestors to shape landscapes to suit their purposes. For example, they could now turn a forest into grassland, a habitat better suited to their hunting skills and attractive to the big herbivores that were the most rewarding prey. Fire also proved useful in keeping large carnivores at bay, especially at night, improving our ancestors' survival chances. Cooking over fire enormously widened the range of possible foods they could digest, improving their nutrition. In short, fire changed our ancestors' place in nature and reduced the chances they might go extinct, as most branches—indeed *all* other branches—of our genus did.

The landscapes our ancestors helped to create in turn altered their genes and their bodies (and thus ours). They became, for example, excellent long-distance walkers, developed arms and shoulders capable of launching projectiles forcefully and accurately, and, because of cooking, stopped investing much scarce energy in building formidable teeth. They developed the capacity to sweat profusely and thereby keep their bodies from overheating during prolonged exercise, a trait shared only with horses.[2] It allowed hunters to run down far faster creatures—antelopes in Africa or kangaroos in Australia—by keeping the prey on the run until it wilted from heat exhaustion. Bodily adaptation to the grassland habitat began in all probability even before hominids harnessed fire, but the use of fire, both to expand grasslands and to cook meat, accelerated and confirmed the process.

A second skill that emerged long ago and influenced our place in nature was the development of complex language. The origins and early evolution of language remain obscure; it may have begun as recently as fifty thousand years ago, although most expert opinion places it much further back in the past. In any case the capacity to coordinate action as it happened, and to plan it in advance, through language, made hunting far more effective and far safer. Language allowed cumulative learning and improved the transmission of information across the generations, a great advantage in competition with other hominids who lacked language and an asset in the struggle for survival against hunger, thirst, and fierce animals. People with language could tell one another where to find water in time of drought or how to avoid certain poisonous plants. This was a great advance over trial and error. Presumably language, like fire, allowed our ancestors to survive longer, to reproduce more abundantly, to secure a larger role in the web of nature, and, once again, to reduce the odds of extinction.

About a hundred thousand years ago, some *Homo sapiens* walked out of Africa and began to spread to other continents. No doubt several groups made the trek at various times. In Africa our remote ancestors had co-evolved over many millennia with various microbes that fed upon them and with the animal prey upon which they fed. This checked hominid biological success. Many microbes eventually "learned" how to use our ancestors' bodies for their energy and sustenance, and to move from one body to the next efficiently, often provoking illness in humans (or hominids). Thus our remote African ancestors lived with a heavy load of parasites and pathogens. Moreover, the animals they hunted had plenty of time to "learn" our ancestors' tricks—more strictly speaking, to evolve adaptations that equipped them to survive in the presence of human (or hominid) hunting.

Once out of Africa, however, our ancestors stole a march on the rest of nature. They left some of their pathogens and parasites behind, bringing many millennia of comparatively good health for the migrants and their descendants. They also walked into landscapes brimming with naive animals that had never been stalked by projectile-throwing upright apes capable of running for hours on end and coordinating their hunting efforts through language. They were, in effect, an exotic invasive species in Asia, Europe, Australia (by perhaps sixty thousand years ago), and the Americas (at least fourteen thousand years ago). As invasive species often do, they flourished in these new landscapes, reproducing prolifically and causing havoc for other species. Thanks to the use of fire and language, and a few stone tools, our ancestors entered new continents as a rogue species, armed and dangerous.

Among the species most affected were the big mammals that provided hunters with the most food for their efforts. Soon after people arrived in Australia, many species of large mammals went extinct. The same thing happened in the Americas.[3] It also happened much later, when people first got to previously uninhabited islands such as Madagascar (about two thousand years ago) and those of New Zealand (about a thousand years ago). These extinctions may or may not have resulted from human hunting; the evidence is usually too slender to permit confident judgment, especially in the Paleolithic cases. Sometimes these waves of extinction are put down to such things as climate changes or epizootics. Climates, however, had changed many times before, sometimes quickly enough to affect species survival. Climate change had nothing to do with the recent extinctions on Madagascar or in New Zealand, but in the Americas the arrival of humans coincided with rapid warming at the end of the last ice age, which might well have helped usher many creatures off the stage. Probably the truth of the matter is that both human predation and climate change were often involved, in different proportions in different settings.

Whatever the constellations of causes may have been, these extinctions had important historical consequences that have reverberated down the millennia to the present. Australia lost all of its largest marsupials, including one the size of a rhinoceros. North America lost giant sloths, mastodons, camels, and horses. This impoverishment of the fauna left human populations in Australia and the Americas with very little to work with in terms of potentially domesticable animals. This situation meant that for millennia Australian aborigines and pre-Columbian Americans enjoyed somewhat better health than African and Eurasian peoples, who suffered from infectious diseases derived from domesticated animals. But it also meant that when strangers arrived on their shores accompanied by horses, cattle, pigs, and a host of unfamiliar infectious diseases, early Australians and Americans were at a profound disadvantage.

About eleven thousand to twelve thousand years ago, perhaps spurred by climate change, people began to produce food by domesticating plants and animals. This transition, long and justly regarded as another of the great turning points in human history (again, not merely environmental history), seems to have occurred first in the foothills of the Zagros and Taurus mountains in the Middle East.[4] However, we probably would not know had it happened earlier somewhere in the tropical rain forests or along coastlines now submerged (sea level was about a hundred meters lower ten thousand years ago because so much water was locked up on land in the form of glacial ice). Archeological remains last much better in dry conditions, such as those of the Middle East, and decay much faster in the humidity of the world's rain forests. So it remains quite possible that earlier transitions to agriculture, or at least to horticulture, took place perhaps in Southeast Asia or West Africa or northern South America. We will probably never know. Agricultural transitions that left some impression in the archeological record seem to have happened independently at least five or seven times, perhaps more, all between twelve thousand and four thousand years ago. These transitions seem to have spread from each point of origin.[5]

Food production allowed much denser human populations than did hunting and foraging. It required a more sedentary lifestyle, even if farmers packed up and moved to new fields every now and then (in a pattern known as "shifting agriculture"). Fields and gardens replaced forest and savanna. When people learned to irrigate crops, agriculture came to involve a new connection to fresh water. Almost everywhere it required dogged labor to carve out fields, to maintain them, to nurture crops, to pluck out weeds, to keep hungry animals from munching the harvest, and in some cases to keep water in the right places at the right times. Soon after people settled down to farming, they began to dot landscapes with villages—the first real "built environment." Of course hunters and foragers had built shelters for themselves from time to time. And probably some fishing communities enjoying especially rich waters built real villages. But only with farming did this become a widespread habit. For all these reasons—the creation of gardens and fields, the careful manipulation of water, and the emergence of the built environment—farming amounted to a thorough revolution in the human environment and the human relationship to nature.

On perhaps a more fundamental level, this transition involved the gradual creation of new breeds of plants and animals—a significant alteration to the biosphere. Over hundreds or thousands of generations, people selected certain plants and animals for preferred characteristics, such as the tameness of wolves (as they became dogs) or abundant seed in the ancient Middle Eastern grass einkorn (as it became wheat). They intervened not with the intent of producing genetic change and new species, but for their own practical purposes which, if sustained over time, inadvertently but inevitably yielded genetic change and new species. Thus thousands of years of more or less consistent human selection turned an unpromising Mesoamerican grass, teosinte, into maize, an amazingly efficient device for turning sunshine into edible calories. Ferocious razorback hogs were turned into docile pigs.

Similar things presumably happened quite accidentally at times. Humans inadvertently, and often counterproductively, selected for certain traits in other species by maintaining certain domestic environments. So, for example, the practice of storing food created new niches for mice, encouraging them to become in effect domesticated, dependent on human communities for their survival. Even insects could become fellow travelers, adapting to human environments. For example, the *Aedes aegypti,* a mosquito native to West and Central Africa, breeds almost exclusively in artificial water containers, pots, buckets, wells, cisterns, or calabashes, rather than puddles, ponds, or lakes like other mosquitoes. The *A. aegypti* also became one of the world's deadliest creatures, serving as the vector for the viruses of both yellow fever and dengue fever.

Viruses and bacteria too adjusted to the new ecological world brought forth by domestication. Most of these adjustments in the microbial world are unknown and inconsequential, but some of them announced themselves forcefully. Bacteria that cycle through our bodies and spread by the oral-fecal route, some of which cause gastrointestinal diseases, enjoyed new scope once people settled down to live in one place day after day and month after month. When farmers came to live cheek by jowl with their livestock (mainly in Eurasia), their health got worse still. Some of the diseases that infected livestock, especially herd animals, evolved into such human diseases as measles, tuberculosis, influenza, and smallpox. Somehow the influenza virus (or viruses, for there have been many) evolved so as to flourish within and among the bodies of pigs, domestic fowl, and humans, leaping from one to the other in the world's farmyards. Perhaps the world's worst human disease, in terms of total lives claimed, is smallpox, which seems to have jumped from camels to humans some time after camel domestication roughly five thousand years ago. Where human populations were dense enough, these diseases could circulate endlessly.[6] Happily, to some extent, the presence of livestock improved human diet, especially where dairy cattle could be raised (and adult human populations developed the capacity to digest milk).

The case of lactose tolerance is another instance of the connections between environmental, cultural, and genetic change, and a consequence of domestication. Lactose is a form of milk sugar. Today all human beings easily digest milk as small children, but most lose the capacity as they age. Before domestication, all people became lactose intolerant after childhood because their bodies stopped making the enzyme that allowed them (as children) to digest it. The domestication of cattle in particular, but sheep and goats as well, offered a tremendous expansion of the food supply—milk, cheese, yogurt—for people able to digest lactose. Sometime around eight thousand years ago a genetic mutation arose in the Middle East or Europe, whereby even adults could digest milk. This proved a tremendous advantage, and by three thousand years ago a quarter of the population in central Europe were milk drinkers.

Today in northern and central Europe 80 percent to 90 percent of the population carries the necessary allele that instructs their bodies to continue producing the necessary enzymes throughout their lives. A similar mutation took place in East Africa, rather later, so that some of the pastoralist peoples there (the Tutsi, for example) are today mostly (90 percent of them) capable of digesting dairy products. Many Fulani—pastoralists of the western and central Sudan—also carry this genetic trait. When and where these mutant milk-drinking human populations arose, which seems to be only twice, they had great incentives to reshape the landscape to make it suitable for cattle, sheep, and goats. In East Africa they tended to do this by fire, controlling the bush and extending pasture in ways practiced by their early human ancestors. In the Middle East and Europe they did this too but also developed agricultural systems in which cattle, sheep, and goats nibbled on stubble after harvests, browsed in forests and on hillsides, and in effect created human food from landscapes humans could not directly exploit for sustenance. In this example domestication led to genetic change within humans that in some parts of the world changed the logic behind land use.[7]

A second example, malaria in Africa, shows some of the negative connections between environmental and genetic change resulting from domestication.

Probably because humans have lived far longer in Africa than anywhere else, Africa evolved a species of anopheles mosquito that drinks only human blood. There are some four hundred species of anopheles in the world, of which about thirty or forty can transmit the plasmodia that cause the human disease of malaria. But most anopheles prefer nonhuman blood and are thus inefficient disease vectors. The *An. gambiae,* however, is the world's deadliest mosquito, because it only bites humans and thus could become a highly efficient malaria vector. Judging from genomic evidence, it did so somewhere around four thousand to eight thousand years ago. The most lethal form of malaria (called *Plasmodium falciparum*) seems to have developed in response to closer-packed human communities in Africa—farming communities—that supported *An. gambiae* in profusion. The disease was so deadly, it provided a rigorous selective pressure for resistance, which meant that genetic mutations that shielded individuals from malaria spread very rapidly through populations. Africans in malarial zones began to develop antimalaria mutations about twenty-five thousand years ago. In Southeast Asia, where malaria also became endemic, different kinds of genetic shields evolved starting around two thousand years ago.[8]

The whole gamut of life-forms, from bacteria to oxen, evolved in new directions as a result of the agricultural transitions, sometimes through enduring human efforts to select useful individuals but more often by chance. And humans themselves changed genetically as a result of new opportunities (the capacity to digest milk) and new risks (the devastating existence of malaria) associated with domestication and farming.

Although human populations grew in number when they shifted to agriculture, they shrank in stature. Skeletal remains show that on average early farmers were shorter than their hunting and foraging ancestors, perhaps by as much as ten centimeters. They ate less varied diets and generally ate less animal protein than did nonfarmers. They suffered more often from vitamin-deficiency diseases.[9] Living sedentary lives among their own wastes, they consequently suffered more from gastrointestinal diseases than did more mobile peoples. Their lives, like their bodies, were shorter than those of hunters and foragers. But they reproduced faster than any other human population.[10]

This exuberant fertility translated into large and dense populations. Combined in many cases with their growing mastery over useful animals—especially horses, camels, and elephants—this population density made agricultural peoples a redoubtable foe for hunters and foragers. Horses, camels, and elephants all became instruments of war. Agriculturalists, especially if they organized themselves into tribes or even states, became militarily formidable and routinely pushed hunters and foragers into the world's deserts, mountains, jungles, and other ecosystems that held no appeal for farmers. Even when their social organization remained modest, farmers reproduced so fast compared with foragers that their sheer demographic weight favored spatial expansion. Pastoralists, who combined mobility with high birth rates and mastery of useful animals, posed at least as strong a threat to hunters and foragers, and drove them from many of Asia's and Africa's steppe and prairie ecosystems (there were no pastoralists in the Americas before 1492 for want of suitable animals).[11]

Horses and war elephants became so valuable for military purposes that rulers tried to govern environments so as to ensure a ready supply of these strategic goods. With respect to horses, this meant expanding or preserving grasslands. In the case of elephants, it meant forest preservation. An ancient Indian author, Kautilya, an adviser to the throne in the Mauryan Empire, explained the importance of maintaining forests as elephant breeding grounds in the interest of the state.[12] Only Asian elephants were domesticated and trained to war; African elephants remained wild animals and from the farmer's point of view, a pest.

City life, which began about 5,500 years ago, marked a another stage in world environmental history. Urban populations generally were so unhealthy that they could not reproduce fast enough to offset their mortality; they were sustained only by constant influx from the surrounding countrysides. Cities were black holes for population until improvements in sanitation and disease control were developed, only about 100 to 120 years ago. The natural decrease (surplus of deaths over births) in London in 1750 was so great, it alone canceled half the natural increase of all of England.[13] Thus urban growth for most of human history was curtailed by the limited supply of potential in-migrants, and big cities could emerge only in the heart of rich agricultural zones or on rivers and coasts well served by cheap transport. Cities in colder climates also made heavy demands on their hinterlands for fuel. According to the calculations of Canadian geographer Vaclav Smil, in the temperate latitudes of Europe or China a city needed to have reliable access to woodlands for fuel amounting to fifty times its own area. In warmer climes this constraint on urban growth did not tell quite so much, because no fuel was needed for heating, but cooking and other uses still required fuel, usually in the form of biomass, often wood or charcoal.[14]

City life also created problems for agriculture. In village settings nutrients from the soil cycled through crops and then human or animal bodies, eventually returning to the soil as excrement, rich in nitrogen, phosphorus, and other elements critical for plant growth. With the rise of urban populations, nutrients often moved from the fields into cities, where they accumulated. Carrying human excrement ("night soil") from cities to the fields compensated somewhat for the long-term drawdown of soil nutrients. But the distance over which such measures were practical was much shorter than the distance over which it was practical to send food to cities. So over time cities reduced the fertility of the fields that supplied them, especially cities perched on riverbanks or seacoasts—usually the bigger cities—because their wastes were often dumped into the water and thus their nitrogen and phosphorus delivered to the oceans and forever lost to farmers' fields.[15] Until the age of chemical fertilizers, the only efficient way to counteract this nutrient loss was the use of manure from animals that grazed in forests or on meadows. They in effect imported nutrients from otherwise little-exploited ecosystems to farmers' fields. Hence, as a sixteenth-century Polish nobleman so pithily put it: "Manure is worth more than a man with a doctorate."[16]

Cities caused problems for agriculture, but they also encouraged its spread and its intensification. Cities became the loci of high-consumption lifestyles (by the standards of the distant past at least), where some people ate more than others, used more flax or cotton, and in general stimulated demand for agricultural goods above and beyond what they could have done as villagers. Greater demand for food and fiber stimulated more agricultural production, which required either bringing more land under cultivation or raising yields—or both. According to a new and controversial hypothesis, expanding agriculture may also have affected climate. Around eight thousand years ago, it seems, the concentration of carbon dioxide (the chief greenhouse gas) in Earth's atmosphere began to climb slowly. This came after about two thousand years of declining CO_2, and when, according to climate models based on earlier alternations of ice ages and interglacials, the CO_2 levels should have continued to fall.

How could this happen? William Ruddiman, an environmental scientist and climate historian, thinks it happened because agriculture spread far and wide. Farmers cut and burned enough forest to send about two hundred billion tons of CO_2 into the atmosphere. This may have forestalled the next ice age. Ruddiman thinks the advent of irrigated rice farming, around five thousand years ago, resulted in extra doses of methane in the atmosphere. Methane is also a greenhouse gas, and its rising concentrations over the past five millennia have helped warm Earth. Ruddiman's views are quite novel and have provoked mixed reactions. If they are correct, they mean that human behavior has been affecting climate in significant ways for eight thousand years, rather than merely in the past two centuries. The emergence of cities, although well after the start of the era of rising CO_2, helped hasten the spread of agriculture and especially its intensive formats, like paddy rice.[17]

For many millennia agriculture remained the most important way in which humankind affected the environment. Agrarian societies outcompeted all others for the most fertile and well-watered lands, continuing to push hunter-foragers, and eventually even pastoralists too, to the margins. Slowly, inexorably, human numbers grew, and more and more land became field, pasture, and garden. Agro-ecosystems spread. Domesticated animal populations flourished. Forest and other wild lands shrank back.[18]

This slow frontier process is the main theme of world environmental history between the emergence of cities (fifty-five hundred years ago) and modern times. John Iliffe, a leading Africanist, made it the central theme of African history in his survey, *Africans: The History of a Continent*, in which Africans are cast as world history's frontier farmers par excellence, struggling to carve their fields from the forests, to keep wild animals at bay, and to keep their populations up in the face of a fierce disease regime. In Africa these struggles often presented stern challenges. Wildlife there posed a more serious threat to human life than anywhere else. The disease climate did as well. This constrained human life directly, via a high human burden of disease, but indirectly as well, because animal diseases (such as nagana, which is spread by the tsetse fly) severely limited the use of domestic animals, especially horses, in Africa. Many parts of Africa had (and have) leached and infertile soils, so African farmers had to learn their soils with great care.

The Sinologist Mark Elvin sees Chinese history in a broadly similar light, as an epic of frontier expansion of the Chinese styles of agriculture, slowly taking over more and more land, assimilating or expelling other peoples, and all the while chiseling the earth into paddies and plots. Hunting and foraging people almost disappeared under this onslaught, driven to the southwestern fringes of China. Pastoral people on the steppelands to the north and west of China proper fared better, mounting a considerable challenge to China from the time of the Han dynasty (second century B.C.E. to second century C.E.) until the armies of the Qing Dynasty subdued the Dzunghar Mongols in the mid-eighteenth century, ending two millennia of competition between Chinese farmers and steppe pastoralists.[19] Elvin has assigned the Chinese state a key role in promoting this frontier process, which is quite different from the African story Iliffe tells. But as ecological phenomena the general patterns are broadly similar.[20]

SAGAS OF EPIC agricultural expansions have also characterized the history of the Indian and European subcontinents, and large parts of the Americas as well. Wherever human populations became large and dense, they did so because of successful agriculture. Large and dense populations (or at least the less well-off people within them) typically felt a need to migrate, to expand, or to set up new colonies. Wherever they had the power to drive off, kill off, or absorb hunter-foragers and pastoralists, they did so. And so, eventually, agriculture covered a third of the earth's land surface, arguably the largest environmental impact (such things cannot be reliably quantified) of the human race.

From the earliest times humans also affected environments by moving plants, animals, and microbes around, both intentionally and accidentally, a process that may be called ecological exchange. Wheat, for example, somehow got from the site of its original domestication, in southwestern Asia, to China by 1500 B.C.E. In times and places where conditions promoted travel and trade (for example, peace), the spread of crops, and presumably weeds and pests as well, accelerated. In the heyday of the Silk Road, for instance, China and the Mediterranean world exchanged numerous useful plants and animals. China acquired grapes, peas, alfalfa, sesame, camels, and donkeys in the era of the Han and Roman Empires (from about 200 B.C.E. to 200 C.E.). Sea routes made ecological exchange feasible over enormous distances. African millets that did well in dry environments were taken to India, expanding the potential of agriculture in the subcontinent; bananas from Southeast Asia came to East Africa, improving the prospects of farming in the moist forest regions around Africa's great lakes. Polynesian seafarers brought a suite of crops and a few animals throughout the South Pacific in their countless colonizing voyages. All this furthered the frontier epics of agricultural expansion; it also promoted a slow process of (partial) ecological homogenization, whereby humankind altered ecosystems so as to raise a handful of rewarding crops.[21]

A famous pulse of ecological exchange followed upon Columbus's voyage from Spain to the Americas in 1492. After the original human invasion of the Americas toward the end of the last ice age, very little interaction took place between the Americas and the rest of the world. The histories of the Western and Eastern Hemispheres, although showing some parallels (such as domestication, state-creation, and urbanism), remained separate. But after 1492, as the eminent environmental historian Alfred Crosby memorably showed, the flora and fauna of the two hemispheres mixed together with tumultuous results. Crosby coined the phrase "the Columbian Exchange" to refer to this ongoing human-assisted migration of plants, animals, and microbes across the Atlantic.[22]

The most disruptive component of the Columbian Exchange was disease. Eurasian and African infections ran rampant among Amerindian populations, reducing them by 50 percent to 90 percent between 1500 and 1650. Amerindians, with their scant experience of domesticated herd animals, had no prior experience of smallpox, mumps, measles, whooping cough, influenza, and several other lethal killers that had become routine endemic diseases in most of Eurasia and Africa. They had no acquired immunities to any of these infections. Moreover, because the entire population of the Americas was descended from a small number of forebears who had migrated across Beringia at the end of the last ice age, these people had minimal genetic diversity among them. This meant that any particular strain of smallpox or influenza that easily circumvented one Amerindian's immune system would likely circumvent everyone's. A more genetically diverse population—Africa's is the world champion in this regard—made the spread of infectious disease more difficult. The catastrophe that befell the Amerindians after 1492 was probably the greatest single demographic disaster in human history, rivaled only by the Black Death of the fourteenth century in Eurasia (and perhaps in Africa).[23]

At the same time as epidemics ravaged Amerindian populations, animals from Afro-Eurasia refashioned American ecosystems. Horses, cattle, pigs, sheep, and goats all crossed the Atlantic with Columbus and many of his successors, finding the Americas a welcoming environment. These animals all ran wild, forming feral herds so large as to amaze observers. They also continued to serve as domestic livestock, first for immigrant Europeans and (mainly enslaved) Africans, and then for Amerindians. These new creatures chewed and trampled their way through ecosystems—including farmers' fields and plots—that were ill adapted to the ways of large hoofed mammals. Their impacts encouraged the spread of immigrant grasses and other vegetation better adapted to life under the hoof. The imported animals also provided American populations, immigrant and Amerindian, with supplies of protein, wool, leather, tallow, and other useful products they otherwise could not have obtained except in tiny quantities (from llamas, deer, and other creatures). As with microbes, rather little of the indigenous American fauna successfully colonized Afro-Eurasia.

When it comes to the plants involved in the Columbian Exchange, however, traffic went both ways. The Americas gave maize, potatoes, cassava, tomatoes, peanuts, and dozens of other cultigens to Afro-Eurasia. Peanut production is so woven into the economy and culture of parts of West Africa that peanut farmers are certain that their crop is native to their part of the world. But it arrived only in the sixteenth century. The potato became at least as central to life in Ireland and the rest of cool and humid northern Europe. Maize became the staff of life in southern Africa, not to mention large parts of the Balkans. Meanwhile several food crops from Africa and Eurasia took root in the Americas. Spaniards and

Portuguese brought vines, olives, and wheat with them. Northern Europeans brought oats and rye. African rice crossed the Atlantic presumably with slave ships and did well in such places as Surinam and South Carolina. Transatlantic migrants brought dozens of other foodstuffs with them, enriching the food supply of the Americas just as American crops did for Afro-Eurasia.

A handful of other historically important plants crossed the Atlantic in the ships of Columbus and his successors. Most of these plants are best classified as drug crops. Sugar and coffee, which were new to the Americas, were the most important economically and socially. They formed a large part of the slave plantation economy so prominent in the history of the Americas from the sixteenth to the nineteenth century. Cacao, another drug crop, went from the Americas to West Africa, where it eventually became a major cash crop.[24]

The economic globalization that followed in the wake of Columbus and other mariners of the fifteenth and sixteenth centuries brought other effects beyond a flurry of ecological exchange. Commodity markets emerged with long-distance reach. The demand for silver in China drove a worldwide mining boom, most rewarding in Japan, Mexico, and the Andes. Mining everywhere changed the face of the earth, spurred deforestation, and, in the case of silver (which was most efficiently separated from its ores by the use of mercury), brought lethal pollution. Fur and hide markets animated a global hunt for beaver, seals, and deer, altering population dynamics and ecosystem balances in northern North America, for example, where beaver had before 1800 played a key role in shaping the landscape (and especially the waterscape). Markets for sugar inspired the creation of a plantation complex, first around Mediterranean shores, then on Atlantic islands, and on the largest scale in northeastern Brazil and the Caribbean lowlands. Sugar meant deforestation, rapid soil nutrient depletion, and biodiversity loss.[25]

THE PROCESS OF economic and ecological globalization lurched into a higher gear around 1500 and is still in motion today. Overlaid upon it, since about 1800, is the emergence of high-energy society based on fossil fuels. The Industrial Revolution is often regarded as a turning point in world history as seen from an economic and social point of view. It is even more clearly a turning point from the point of view of environmental history. Before the harnessing of fossil fuels, people had great difficulty deploying energy in anything but tiny amounts. The main way to do it was through human muscle power, supplemented in cases with animal muscle and in a few select locations, wind or waterpower. Most everything, from building pyramids to carrying freight, required muscle power. This was a great constraint on how much work could be done, and therefore how much wealth might be created. It also helped to account for the widespread practice of slavery in the preindustrial era, as there was no more efficient way to get big things done than to amass human muscle.

Fossil fuels changed all that. They represent a subsidy from the geologic past, bestowed upon the past six or seven human generations (and probably the next several as well). Their ecological effects were, and remain, enormous. Fossil fuels made the big cities of the industrial era by snapping former constraints on urban growth. They allowed enough food to be brought in fast enough to keep millions of people alive. They eliminated the need to bring vast quantities of firewood into cities each day. They made the factory labor of the toiling masses so much more productive that factory owners, and eventually laborers, could afford to consume cotton, tea, sugar, flour, and other products brought from far continents, changing the landscapes of India, Egypt, the Caribbean, and Australia. Initially, in the first two generations, industrial laborers were more malnourished than their country cousins. Like the Neolithic Revolution before it, the Industrial Revolution at first made human beings shorter in stature.

Fossil fuels were a dirty innovation. The first industrial cities, in Britain, were horribly polluted as a result of the burning of coal.[26] Children living in these cities developed rickets, a result of vitamin D deficiency because so little sunshine reached their skins (where healthy children manufacture vitamin D). Everyone breathed quantities of particulate matter as well as chemical pollutants released in the combustion of coal. As a result, lung disease rates soared. Mining coal was a messy business itself. It filled local environments, and miners' lungs, with coal dust, which gave millions of them emphysema and other lung diseases. Mining also dotted landscapes with slag heaps. Eventually, by the late twentieth century, more technologically sophisticated coal-mining methods involved lopping the tops of mountains and filling valleys and streambeds with rubble and slag.

The use of coal had important indirect ecological effects, mainly through cheap transportation. Steamships and railroads opened up broad new landscapes to systematic economic use, changing landscapes and ecologies wherever they went. Refrigerated railroads and steamships by the 1880s made it practical to turn large swathes of the United States, Canada, Argentina, and Australia—a goodly portion of the world's grasslands—over to meat production. Coal also helped inspire vast increase in the mining of such useful metals as iron, tin, nickel, and copper, because metallurgy based on a cheap and abundant fuel was so much more profitable than metallurgy based on charcoal. Without the cheap energy coal embodied, Chile's copper, Malaya's tin, Ontario's nickel, and Australia's iron ore would still lie underground.

Oil made energy cheaper than ever before. The first major strikes occurred in the 1850s in Pennsylvania. Great gushers followed, in Baku, Austrian Galicia, Texas, California, Venezuela, Mexico, Iraq, and Sumatra—all before 1940. Then, in the late 1940s, petroleum engineers developed the great underground sea of oil around the Persian Gulf. The second-largest deposits in the world, in western Siberia, opened up in the 1960s. Drilling, transporting, refining, and burning oil were all dirty businesses with widespread consequences in the form of spills and pollution of air and water. But the largest environmental consequences of oil, as with coal, came not from its direct pollution effects, considerable though they were. Instead, they lay with the economic activities that oil made possible. In combination with new machinery, cheap oil made it economically practical to tear the top off of mountains in search of a few grams of gold. Cheap oil (and chain saws) made possible a sudden spurt of cutting and burning in the world's tropical rain forests after 1960, a major ecological change of our times, and one that could not have happened quickly without oil. Cheap oil also underlay the rise of the personal automobile, car culture, and the consequent changes in urban air pollution, land use, and city layouts.

The power of cheap energy to enable sweeping ecological changes, to make things happen faster and more broadly than they otherwise could, reached every corner of the globe. Cheap oil made the fertilizers and pesticides of modern agriculture feasible, as well as the agricultural machinery and transport networks that help bring food from fields to tables almost everywhere. Without it the yields of agriculture would be roughly half of what they are, and the quadrupling of human population since 1910 could not have occurred. Thus one can logically say that oil has doubled our food supply and population, and all the environmental consequences that flow from having six billion people on earth (rather than three billion) are owing to oil.

Fossil fuels, both coal and oil, may constitute a turning point in environmental history in yet another way. They currently account for upwards of three-fourths of the greenhouse gas loading of the atmosphere that is warming the planet. Historically and cumulatively, they probably account for somewhat less, but surely more than half of the greenhouse gas emissions. If global warming turns out to be a severe challenge to human habits, or to require major adjustments in patterns of life, then the advent of fossil fuels will qualify still more convincingly as a turning point.[27]

Over the past hundred thousand years seven turning points in world environmental history stand out: (1) the domestication of fire, (2) the emergence of language, (3) the exodus from Africa, (4) the domestication of plants and animals, (5) the rise of urbanism, (6) the Columbian Exchange, and (7) the adoption of fossil fuels. Most are widely recognized as turning points in history in general, and all should be. All represent basic renegotiations of the human place within the biosphere. They came at faster intervals in more recent millennia, a result of our species' growing power and numbers. In the fullness of time there will be other turning points, the nature of which we cannot yet guess.

3. Agriculture

DEBORAH FITZGERALD

Nearly every important change in agriculture has been caused by (or caused itself) an important change in the environment. Agriculture and nature are so inextricably bound that a perturbation in one means a shifting in the other. There are some dramatic examples of this: for example, three-field rotation during the medieval period, the British enclosures, the Caribbean sugar industry, the introduction of sheep into Mexico, and the American Dust Bowl. Drastic changes like these, as well as more diffused ones—such as farmers shifting from diversified farming to monocropping—have had notable consequences. Since about 1850, many of these changes have been driven by scientific and technological innovations, which in turn has increased the pace and often the costs of such changes. All of these changes have been at the center of new social relations of production and consumption.[1]

But what makes a big change a turning point? This is a knottier question than it might seem. If one wants to avoid utter reductionism, as I do, one must recognize that not all changes have a big impact unless and until they become widespread. A turning point needs momentum. In general, turning points have been occasioned by events and processes that have increased the scale of production, but these often take time to work their way into common practice. A variety of such processes might be at work. Opening new lands to production, tools and machines that allow people to tend more land, bringing new animals into the farming scheme, wartime demands and opportunities, and innovations in transportation have all led to substantial shifts in scale. But each of these changes might be carried out by individuals widely separated in time and space, each having minimal impact on the larger trajectory. The invention of the stationary steam engine could not itself occasion a turning point unless a lot of people put such a machine to continued use in a fairly large area. The development of new genetic strains cannot occasion a turning point if most farmers cannot afford to buy them, or if they require other, largely unavailable inputs to function well. In addition, few big changes in agriculture have been caused by a single person, technique, or innovation. Nearly all such changes have depended on the combination of things—tools with capital, expertise with state power, demographic shifts with transportation, and crops with diet.

For the sake of argument, following from these observations, we can suggest that the real turning points in agriculture collect around three historical periods. The actors have been western Europeans, Americans, and enslaved Africans and Caribes. These revolutions have been marked by dramatic increases in productivity, albeit from different causes. The first such revolution was the mid-sixteenth to mid-nineteenth century, the second from the mid-nineteenth century until World War II, and the third from World War II to the present. These are fairly sloppy boundaries, with much leakage from one time frame to another, especially when considering the international context. But this offers a plausible framework. This chronological periodization somewhat obscures the equally important causal dimension: What makes a time frame a turning point? Can we talk about turning points that affect only a small part of the world, or can a turning point be localized? Plant genetics, for example, might well be considered a turning point, but it is a long one. Mendelism was rediscovered in 1900, yet many parts of the world are just hearing about Mendelian genetics and joining the shift from traditional crops to genetically modified ones. These two kinds of change must be kept in constant relation if we are to understand agricultural and environmental transformations.

Revolutions and Turning Points

The first agricultural revolution in modern times occurred in Great Britain and featured two distinct but related features. One strand, known by British historians as "the agricultural revolution," stretched from between 1500 and about 1750 and centered on the increased use of livestock, the resulting increase in manure to fertilize fields, the growing of root crops, and the enclosing of open land. This gave farmers and landowners more flexibility and control over both crops and livestock. The combination of better soil and more plentiful livestock stabilized agriculture and rural life at a time of tremendous transformation. It is quite striking, when looking at a period before industrialization, that rural people shifted their farm operations in line with current and emerging circumstances. Climatic changes, market ups and downs, labor availability, large demographic changes, and foreign influences could all have an important and relatively immediate impact on rural practices. The environment and the agricultural community were in constant discourse, by necessity. Today, the logic of agriculture tends toward the technological fix—cold winters but plentiful energy supplies and plentiful capital lead to greenhouse vegetables year-round, for example, and market orders necessitate shipping produce to cold climates from warmer ones all year long. The link between agriculture and environment has become somewhat attenuated for us, contributing to our lessening consciousness of it.[2]

The second strand of this "revolution" was the development of plantations in British colonies, particularly the Caribbean islands. As Sidney Mintz and other scholars have explained, the rise of the "sugar colonies," such as Barbados, marked the beginning of a large-scale but diffuse system of production. Discovering a climate favorable to sugar cane and enslaving an indigenous population, British and later Dutch colonizers leveled the lush island forests and built cane fields on the often precipitous hillsides of the islands. They built factories for squeezing the sugar out of the cane and turning it into brown sugar or rum. Thus were combined massive alterations to the landscape, the forced labor of indigenous populations, elaborate ocean trade, and the creation of a food that arguably underwrote the Industrial Revolution.

One could argue that this colonial process prefigured the American slave trade and its cultivation of tobacco and cotton in the southern states. Here again, a planter class acquired African and Caribbean slaves to plant and cultivate a commodity for export—a circumstance that led to a pivotal American dislocation, the American Civil War. Abhorrent as it is to us today, plantation slavery was a powerful form of agricultural practice, debilitating for slaves and the land alike. In the southern states the constant cropping of land in cotton and tobacco, and the lack of any practices to build the soil back to productivity, created a historically poor agrarian underclass in the postslavery era, as they tried to raise subsistence crops on devastated land.[3]

The second agricultural revolution, from the mid-nineteenth century to World War II, featured the application of science and technology to production and, to some extent, consumption. To my mind, this was by far the most explosive and dramatic turning point, the main features of which were in place by about 1930. There were at least four strands to this revolution, each of them fundamental in explaining what happened over the next hundred years but quite different in terms of the pace and depth of change. The first strand was the development of agricultural science. The second was the organization and institutionalization of agricultural knowledge, especially in America. The third was the application of machinery to the landscape, particularly the development of the railroad but also the internal combustion engine. The fourth was Mendelism and greatly accelerated plant and animal breeding in general. Each of these innovations caused a distinctive uptick in agricultural productivity but also introduced new experts, rules of practice, and institutional homes and protocols for agricultural activities.

It may seem obvious that the systematic collection of known facts about a place—the identification of different climates, degrees of aridity or rainfall or soil type, the ease of growing certain kinds of crops or raising certain kinds of animals—was a crucial stage in the development of modern agriculture. What may be less obvious is that people did not always keep this kind of information. Its collection varied greatly from place to place, and often local people followed habit and tradition even when they were at odds with experimental results. Furthermore, what was true in one country might be quite different in a neighboring one, so it was always hard to achieve an accurate and complete understanding of the facts of any place bigger than a township. As the anthropologist James Scott has suggested, in truth only the state really cared about the larger canvas, in part because it was in the state's interest to know for certain how much food and livestock could be requisitioned in times of strife.[4]

The collection of agricultural facts tended to be supported by the state wherever it occurred, and the state also paid for the creation of recommendations and standards in agricultural practice and trade. One of the most important efforts along these lines was the development of the science of soil chemistry, in Germany in the nineteenth century. Based on an experimental and time-consuming method, Justus von Liebig was able to figure out how to test the nutrient content and structure of the soil and to correlate that to different planting requirements. His role was so important that young American scientists who studied with him brought back to America not just the science but also the academic organization that made it possible. Soil science was the primary scientific focus in the new agricultural colleges, and the experimental basis upon which it was based lay the foundation for succeeding sciences, such as plant breeding, plant pathology, and horticulture. For most of these scientists the American landscape was a rich source of sustenance and a place filled with challenges to healthy agriculture in the form of wild weather, nasty insects, and uncertain soil quality. For early farmers the careful nurturance of the farm by many was offset by the thoughtless disregard by others, who chose to keep moving west and south rather than deal with indefinite results.[5]

Many scholars underestimate the role of agricultural institutionalization in accounting for the growth of agriculture in the late nineteenth and twentieth centuries, but in fact it was of paramount importance. In the United States this agricultural institutionalization began with the Homestead Act in 1862, which threw the entire country open to settlement by anyone who would promise to develop forty acres of land. Despite the fact that the railroads and speculators managed to acquire some of the choicest land for themselves, the Homestead Act moved millions of new immigrants out onto the prairies and into the western lands to begin farming. Some of this land was ill-suited to the plow, or even to grazing cattle and sheep, but this mismatch only hastened the development of technologies to irrigate and subjugate the arid lands. Also in 1862 the Morrill Land Grant Act promised every state in the Union enough land to build a college of agriculture and mechanic arts, a democratic scheme of unprecedented reach. Within the next few decades the forty-eight states each built an agricultural college and hired the scientists, engineers, social scientists, and humanists to train rural youth in not only agricultural knowledge but the wider world. At the same time the United States Department of Agriculture (USDA) was established in the nation's capital, with the aim of encouraging the development of agricultural knowledge and sound practice.

Over time the USDA grew into a massive government organization, responsible not only for research but also for the regulation of all agricultural practices, agricultural and home economics education in all of the states, and a good deal of federal legislation. By 1887 the colleges were up and running, for the most part, and farmers were becoming enamored of the expert opinions that college professors provided. The Hatch Act was enacted to provide funds for research into agriculture at a time when professors were having trouble finding time to conduct it. In 1906 the Adams Act continued this arrangement. In both of these cases the federal government was encouraging the scientists and engineers to get a handle on the agricultural facts of their states: What kind of soil did it have? What kinds of crops and livestock did well there, and which did not? How could farmers make a good living from the land? How could farm homes be made more congenial? The government was also supporting scientists and engineers in pursuing original research that would improve rural lives and pocketbooks. Just over fifty years after this agricultural machine had gotten started, the Smith-Lever Act was passed. Through it the federal government gave money to the states to "extend" this new information on agriculture and rural betterment out into the farming communities, bringing the original institutional design full circle.⁶

Although it may seem odd to count the development of this institution as a revolution or turning point, the size, reach, and legal authority of the USDA created a momentum unmatched by any other body. This was important particularly because the department made policy that directly shaped environmental circumstances in the states. For example, the USDA developed policies relating to settlements in the arid West, allowed ranchers to graze cattle for no cost in the West, offered subsidies to farmers who operated in flood plains (enabling them to continue farming there despite repeated floods), created policies during the New Deal to encourage some farmers to grow more and some to grow less, generating in one case a program that pays farmers to *not* produce some crops in some areas, and so on. Certainly when American historians think of environment and the state they tend to think of the Bureau of Reclamation and the Department of the Interior, but in fact the USDA did far more to dictate the contours of the working environment thought of as "farms."

Yet another feature of the institutionalization of agriculture was the emergence of a large group of agricultural experts, people whose advanced degrees in the new agricultural fields of agronomy, agricultural engineering, home economics, agricultural economics, and veterinary science (to name just a few specialties) created an unprecedented divide between those who studied agriculture and those who practiced it, and between those who had authority and power and those who did not. For many farmers these experts were a godsend, because they could help farmers figure out why things were not going right on the farm, and they could keep farmers abreast of new developments, techniques, and applications. The extension agents were especially useful to farmers. Trained in a specialized field, and employed by the agricultural college explicitly to help farmers who needed it, the agents were familiar with both local conditions and national trends. At the same time, however, agents and other agricultural experts could become impatient with "traditional" ways of farming, always urging farmers to upgrade their tools, machines, and seeds, or to adopt expensive innovations.

Often their advice was very sound and helped hide-bound farmers make the transition to more sensible practices. But just as easily the experts could overlook traditional practices that made local, if not national, sense, or that solved other problems that the agent did not appreciate. The biggest example of this was the shift from diversified farming to monocropping. Again, this made sense if all the farmer wanted to do was to expand his acreage and focus on one kind of market. But it did not make sense if the farmer wanted to spread his risks over a variety of operations (for example, small grains, orchard fruits, livestock, home garden, chickens, egg or milk production, and so on). This was also a popular style of farming if the family wanted to produce its own food cheaply and had the children do much of the work.[7]

The third strand, mechanization, refers to the application of power to the landscape. Here I mean not only tractors and steam engines but also, and perhaps more important, the railroad. The building of the railroads that crisscross the country, and especially the completion of the transcontinental railroad in 1869, changed people's relation both to the market and to their own communities in profound ways. Before railroads, farmers were limited in what they were able to send to high-paying urban markets. Animals had to walk there, and foods and grains had to be carted at considerable expense, quickly for things that could rot. Dairy products had a very short life before refrigerated railcars were developed, and even then they were prone to spoilage. Similarly, rural people were isolated from and, depending on your point of view, protected against the splendors of urban life. Mail, manufactured goods, and visitors were able to go to the countryside only with difficulty and patience.

The railroad changed all that, although not all at once. But for those who could see the train's promise, the payoffs were handsome. One of the most strategic groups were the fruit growers in California. Not to be confused with the farmers of Jeffersonian myth, the California fruit growers were marketers of fruits and vegetables grown in a climate unusually perfect for that purpose and unique in America. Their success in growing fruit in particular became problematic because the best markets were on the East Coast, three thousand miles away. The railroad gave these growers the means to get their highly perishable produce to large and well-paying markets year-round. In turn, this gave eastern consumers a more varied diet. Wheat growers in the West and mountain states likewise found the railroad an essential component of their agricultural strategy. Indeed, some rail lines persuaded immigrants and homesteaders to settle in western areas, where the train passed through in hopes of developing a grain trade to support the eastbound trains.[8]

The other kind of mechanization occurred in the field and the farm home. The first sort was the stationary steam engine, a huge behemoth that powered wheat-harvesting machines. In distinction to most of western Europe, the American landscape features many large, flat, uninterrupted areas that farmers used for grains and cattle. These large grain farms were legendary for their sheer scale. For example, the "bonanza farms" of the Red River Valley of the upper Midwest drew visitors from all over the world interested in seeing the machines bring in the wheat. Inventor Cyrus McCormick's reaper was justly famous in the wheat fields because it allowed farmers to employ fewer workers at the harvest, while operating more acres. But it was the internal combustion engine that really revolutionized fieldwork and farmwork alike. The internal combustion engine made possible automobiles, trucks, tractors, and combines (that is, "combined" harvester and thresher machines). Automobiles were important not only for enabling farmers to travel off-farm to pick up supplies, deliver goods, and so forth, but the early Ford engines could also be isolated as power drives to run other small machines, such as grinders, washing machines, and a host of other handy tools. Trucks were used for taking produce and livestock to market and for hauling around farm-related things like fencing, livestock feed, tools, and hired help.

Both cars and trucks were available to farm families by about 1915. Tractors were also available, but reliable ones were not common until about 1920. Tractors and combines are often seen as labor-saving, although this is probably not quite accurate. Both machines tended to replace not hired men but horses that actually did the fieldwork, such as cultivators, seeders, weeders, and so on. Most early machines were too expensive and too temperamental to trust to hired men; it took a fair amount of skill to operate these machines, and schools cropped up all over the country teaching young men how to do it properly. Combines became available in the late 1920s, which made it possible for farmers to manage much bigger farms than ever before. Wheat required little attention between planting and harvest, but at harvest time it had to be brought in quickly. This time pressure prevented most farmers from planting large wheat acreages until the combine appeared. It was a double-edged sword, however. Farmers were able to expand their farms by using combines, but they also had to expand to pay for the machine, which was expensive by any standard. Thus began what one economist called "the agricultural treadmill": farmers expanded to pay for the means of production, which required them to expand.[9]

These machines had a significant impact on the landscape. For one thing they encouraged the shift from diversified family farms (about 160 acres) to large, single-crop farms (more than 800 acres). Where diversified farms tended to reward attentiveness to the soil and landscape, monocropped farms rewarded scale and speed. Where diversified farms tended to support those who lived on the farms with food as well as commodities, monocropped farms produced only commodities and often no-one lived on the farm itself. In the most extreme cases workers went to the farm as they would have gone to the factory; it was certainly not their home. Diversified farms tended to contain trees, creeks, paths, woods, and meadows, while monocropped farms tended to be a solid block of grain without any other natural features. Visually a field of wheat could be quite beautiful, even awe-inspiring. But this beauty hid the truth that such cropping was limited because it put nothing back into the soil; wheat farmers did not usually rotate crops or do much to build the soil. The result was "thin" soil—soil that lacked nutrients and structure, the kind that blew away in the Depression-era dust storms. Land that had no trees or creeks or woods could not break the wind and could not hold the soil: the legacy of mechanization gone bad.[10]

A related problem was the erosion of land. Certainly this problem could not be attributed entirely to agricultural practices. As filmmaker Pare Lorentz's dramatic propaganda film *The River* points out, only somewhat hyperbolically, land that was deforested to build the cities ultimately ended up downriver, helpless against heavy rains and storms that pushed it into the rivers and downstream. Yet farming was a large part of this deforestation in many parts of the United States, as farmers cleared land for fields with little regard for the land's contours and relation to waterways. The Civilian Conservation Corps responded to this crisis in the 1930s by planting trees, including "shelter belts" in the plains, and by showing farmers how to terrace their fields for maximum soil conservation. One thing that became clear was that family farmers were much more attentive to this message than investment farmers.[11]

Genetics and Biotechnology

The final strand of the scientific and technological revolution was genetics and biotechnology. These may seem to have little to do with environmental issues, but they were quite importantly intertwined by the 1920s. Genetics dates from 1900, when Mendel's experiments were discovered, and private breeders as well as the USDA began using Mendelian principles in breeding maize, the first agricultural crop to be hybridized. Genetics was attractive to breeders for many reasons. It seemed to promise to breeders something akin to patent protection. Breeders felt that they invested a lot of time and money in developing new strains only to watch farmers ruin them (the patent idea fell apart in the face of serious legal and biological realities but returned with a vengeance in the 1970s).

Breeders also were excited to be able to breed plants for specific features, such as drought resistance or heavier ears. Hybrids were the first new thing to come along in agriculture in a while, and farmers would be willing to pay more for seed that produced larger crops. (Indeed, they had to pay three times as much.) But as with so many new technologies, the hybrids did not stand alone. They required a host of other, complementary technologies and practices as well to work at all. Farmers who planted hybrids without adequate rainfall or irrigation, or herbicides and pesticides, or special fertilizers, could well end up with less than they had begun with. Hybrids tended to be difficult to deal with unless the farmer bought into all the other "inputs" that the breeder or salesman suggested. The upshot of all this was that farmers started putting lots more chemicals on their fields that they ever had before, a trend that grew as the century progressed.[12]

The third revolution, from World War II to the present, is in some ways a larger version of the second revolution but with some twists. The main feature of this period has been the speed and near-totality with which agriculture has become industrialized in the West, something that is strongly tied up with the increased demands of the war itself and the new scientific and technological capabilities developed to solve military problems. Agriculture has become much more "input" oriented (fertilizers, pesticides, herbicides, irrigation), diversified farming has given way to specialized farming, mechanization has become the dominant form of farm power, and the number of farmers has decreased precipitously. The current issues raised by emerging biotechnologies are in many respects old issues that have been accelerated.

Wartime has always provoked crises in food production, distribution, and preservation, and these problems have in turn led to changes in farming practices. For example, much of what soldiers have needed most has been food that could be preserved over a long period of time. Although fresh food might be what soldiers craved, it was impractical to depend on it during a military campaign. Modern food processing has been closely tied to war-related demands. Food canning dates from the Napoleonic War in France and from the Civil War in America. Such dried foods as eggs and milk were developed during World War I, and in World War II fruits, vegetables, and meats were also dried. Soldiers could reconstitute these foods with water in the field, and mess tent and submarine cooks depended heavily on such space-saving foods.

In general, the wars increased the number of acres of farmland in production as well as the number of meat and dairy animals. This could lead to unusual problems. In California, for example, processors of pears and asparagus were in the habit of throwing the peels, seeds, and stalks into the river, using only the choicest part of the fruit or vegetable for processing. Fruit growers in Oregon and Washington also disposed of peelings in the rivers. Sport fishermen were the first to realize this was a problem, discovering that it was killing the fish and destroying the river's ecosystem. Processors soon had to find other ways to dispose of their garbage, sometimes plowing it into fields instead.[13]

The way that farmers operated fundamentally changed during World War II. Using Iowa as an example, one can see how the war changed both what farmers produced and how they produced it. Before the war Iowans were nearly self-sufficient in food production. Farmers grew not only commodity crops such as maize, alfalfa, and soybeans, but also a wide variety of fruits and vegetables. Most Iowa farms kept chickens for both meat and eggs, and many kept a few cows for milk. Most also had hogs. Farms traditionally included large kitchen gardens, which provided the farm family with both fresh and canned produce throughout the year. Many farmers grew fruits and vegetables for processing as well as for the market; before the war there were around sixty canneries scattered around the state that depended on Iowa farmers for seasonal produce.[14]

During the war itself not much changed in Iowa agriculture, except that farmers produced more than ever before. But changes in other parts of the country signaled shifts ahead for Iowans. One of the biggest changes was the explosive growth of California fruit and vegetable growers. Although California had long been a force in the canned food trades, its role in fresh produce expanded greatly during the war. So much so that after the war it was cheaper for Iowans to buy fresh produce from California than from local growers. This phenomenon was part of a larger rearrangement of the geography of food provisioning in America, characterized by concentrations of food production in particular places (for example, fruits and vegetables in California), more standardized grades and types of food, and decreasing shipping rates that made it cheap to transport foods around the country. In the 1950s the emergence of supermarkets and the interstate highway system accelerated these trends. The chicken industry also grew out of this trend toward concentration. By war's end Iowans had begun to stop keeping chickens for food (although they continued with eggs for a time) and instead started buying chickens in the stores year-round. The industrialization of chickens as well as egg production, and their concentration in the southern states and the Delmarva Peninsula (Delaware and portions of Maryland and Virginia), spelled the end of casual chicken production in Iowa.[15]

One by one the diverse pursuits of traditional farms gave way to the primary focus on commodity crops that one sees on most American farms today. In most cases there was an economic reason for the change, as farm families experienced the same modernizing momentum as others after the war. Expanded acreages required machines to ensure speedy harvesting and cultivating, and paying the high price of these machines led farmers to grow crops with a guaranteed market and federal subsidies. As more farm women took jobs in town to supplement their family's income, women's labor was no longer available to tend chickens, calves, gardens, and hives. For most families these changes were gradual and barely visible during the transition. But a before-and-after photograph of the farm would tell the tale of changing land-use patterns, scale of agricultural practice, and shifting priorities.[16]

For some agricultural experts in Iowa the postwar economic slump in commodity prices suggested that farmers might do well to get out of commodity production and into the production of what they called "high value" crops, such as fruits and vegetables, honey, eggs, and chickens. They organized several large buyers and processors (for example, Heinz) to pledge their support for Iowa vegetable crops and began experimenting with different varieties and cropping plans. But they ran up against an unexpected problem. The chemical agents that had made fighting in jungles possible—DDT and 2,4-D, to name the most important ones—now were being promoted to farmers to reduce their weeds and crop pests and to clear ditches of weeds when the highways were built. Applied by airplane, the chemical clouds drifted far beyond their target field, laying poison indiscriminately on less hardy plants and animals. Vegetable growers found that their plants would not grow, and fruit would not bud. Without the wild clover in fields and ditches, bee populations declined to almost nothing, further damaging the fruit crop and severely hurting the Iowa honey industry, one of the largest in the country.[17]

Other crops have also succumbed to the geneticist's desiderata. The tomato, for example, was transformed not only for optimum flavor characteristics (which genetics has done little to improve), but mostly for its picking and packing characteristics. Its development occurred in tandem with the creation, at the University of California at Davis, of a tomato-picking machine, designed to address an impending labor shortage (which did not occur). Although engineers could design a good machine for picking tomatoes, none of the existing field tomatoes could withstand the machine's metal "fingers," which destroyed the delicate fruits. Plant breeders decided to develop a new tomato that had tougher skin and a more square shape: easier to pick and to pack in crates. The resulting tomato-picking machine combination led to a lawsuit against the University by California Rural Legal Assistance, which charged the university with abnegating the provisions of the Hatch Act. The act had given the university research funding to improve agriculture for the state's citizens, the plaintiffs argued, not only a few wealthy growers.

The lawsuit was initially successful but then lost on appeal. But the early result of the tomato-picker combination was that medium-sized tomato growers, who grew tomatoes in rotation with other vegetables, got out of the business because the picking machine was too expensive for them. They could not compete with the big growers and packers without it. One of the environmental consequences of all this was a much more brutal picking situation, with the whole tomato plant now yanked out of the ground as a multiton machine rolls its way through the field. An industrial crop, tomatoes were also dependent on a steady application of agricultural chemicals—herbicides, fertilizers, and pesticides.[18]

In response to the extraordinary scale of agricultural chemical use, and its disastrous effect on groundwater supplies, some proponents of today's new biotechnologies have argued that the real virtue of modified crops (such as those with the Bt gene) is that all the insecticide is inside the plant. The insect pest therefore dies by chewing on the plant, not by getting doused with pesticide. But critics have charged that we still do not know enough about the long-term effects of this kind of genetic manipulation. One problematic study suggested that Bt corn crops had a deleterious effect on Monarch butterflies, a finding that played havoc with American corn contracts both at home and abroad. International bans on American grains continue to reflect the widespread concern that no one really knows what these genes will do to other biological systems—whether human, animal, or plant.[19]

Increasingly it is clear that institutional momentum is extremely important in the international agricultural context. If the primary indication of a turning point is the sustained and widespread adoption of production-increasing ideas and innovations, then surely the post–World War II spread of agricultural science and technology has been an important story. Here the burden of change lies not only with the state, although the state has been crucial. In international agriculture the World Bank and especially its Consultative Group on International Agricultural Research (CGIAR) has been at the forefront of "modernizing" international agricultural practices. Their network of research centers and extension agents have in some cases shored up traditional practices, but more often promoted the adoption of scientific and technological innovations, such as hybrid seeds and tissue culture. But multinational agricultural companies like Monsanto and Archer Daniels Midland, which have their own research scientists and engineers, have gained a substantial foothold in the international markets, exerting considerable pressure on village farmers to adopt a packet of improvements that might include seeds, fertilizers, irrigation technology, and marketing. This is not entirely novel. Indeed, early machine manufacturers such as International Harvester and McCormick traveled the world in the 1920s selling their equipment and establishing sales agents. But the scale did not approach contemporary efforts. Today's corporations seem committed to international markets as a critical venue for agricultural innovation and sales.[20]

WHEN WE REFLECT on big environmental issues having to do with agriculture, most of us think primarily of agricultural chemicals—pesticides and herbicides, and large implements (think: "the plow that *broke* the plains"). These innovations did indeed have a huge and evidently lasting impact on both agriculture and nature. But they must be understood in the context of institutions, state and corporate organizations that form the grid work on which these scattered and uneven innovations hang. To ignore the institutions is to miss the complicated way that farmers can become caught in a web of much larger meaning and consequence than their individual actions would suggest.

In considering turning points, let us consider Joan Thirsk's argument for "alternative agriculture."[21] The notion of revolutionary changes or turning points in agriculture projects a linear time line in which one thing leads to another new thing. Yet, as Thirsk has pointed out, agriculture can also be viewed as cyclical, particularly when taking the long view. Although farmers have traditionally focused their efforts on meat and grain, they have also sometimes been forced to seek alternative crops, livestock, forms of income, and food. When they later return to the mainstream (often a century or more later), they bring something of the alternative experience with them. How, then, might these interludes of alternative agriculture map onto the larger agricultural canvas—land, water, countryside?

A modest example of this might be the current fascination with organic crops or with local foods. "Organic" and "local" are in some respects just another way of saying "before commercial inputs were developed." Organic fruit is fruit that is not sprayed with chemicals or often not bioengineered with another organism's genes. What is this but a return to a preindustrial way of producing foods? Similarly, "eating locally," now considered a mantra of privilege, was once the cry of the deprived, those who were unable to eat foods they did not grow themselves or to use things not available in their own country (for example, coffee, chocolate, spices). These movements are if not backward-looking then at least a part of a longer cycle of surplus and want, extravagance and penury.

Finally, one often overlooked but important element of agriculture (one that has substantial implications for environmental history) has been settlement and unsettlement. In America, for instance, the introduction of European agricultural practices, as well as the westward movement of agriculturalists, left noteworthy imprints on the land. The use of suboptimal land for agriculture led to the Dust Bowl and the draining of many western rivers and lakes, which in turn led people to populate new areas and depopulate old ones. Internationally, how have shifting populations, and their accompanying agricultural practices, led to environmental improvement and/or destruction? Does settlement invariably lead to declining environmental circumstances, and does unsettlement lead to an improved ecology? This is the kind of question that can only be addressed, perhaps, by the collaboration of geographers, historians, and naturalists, but it is worth considering. Migrations on both a small and a large scale have played a huge role in global environmental histories, and by linking these stories to agriculture we can develop a much more robust framework for understanding the real turning points of our lives.[22]

4 Forest History

BERND-STEFAN GREWE

From the battlements of the former castle near Hambach, Germany, a one-hour walk from Neustadt, there is a wonderful view over the Rhine Valley—about twenty kilometers north, forty kilometers east, and up to eighty kilometers south. If the weather is clear, the cities of Ludwigshafen-Mannheim, Heidelberg, and Karlsruhe are visible; the low mountain range of Odenwald and the northern part of the Black Forest can be seen too. Perhaps this is the reason a castle was constructed on this wooded hill in front of the mountains some nine hundred years ago. The scenery makes it an attractive place for an outing. On many weekends the car parks close to the castle are overcrowded with daytrippers from Ludwigshafen-Mannheim or Karlsruhe. Nowadays, forests are a space of leisure, where one meets hikers and other city-people doing Nordic walking, jogging, or mountain biking. Upon entering, one sees dark and dense forests that consist of only one or two different species of trees, all of the same age and size. There are very few clearings, with the exception of crossroads where truck-trailers can load timber.

For many centuries this was not the case. The castle was not in the woodlands yet, but instead some hundred meters away from the trees. The woods were not very dense and had many clearings. Walking on small paths through the woodlands in that time, one would have seen cattle and horses grazing between the trees and pigs searching for acorns. Some parts of the forest would have been cultivated for one or two years with wheat and then been left to the trees again. Many women and their children would have been observed cutting grass or collecting leaves from the branches to feed their livestock at home. One might have met some strange and wild-looking people with black faces—these being the charcoal and potash burners living inside the forests.

In spring 1525 the idyllic view from the battlements at Hambach Castle became threatening when some thousand peasants successfully attacked the castle. In addition to political rights and some religious liberties, the peasants also demanded back the common forests, which the territorial rulers had appropriated. In vain. The rulers united their armed forces and struck back, and shortly after the Peasant War was brought to a dreadful end. This was undeniably a turning point in German history, as the dominion of the nobility persisted for another three centuries. The rulers kept their grasp on the forests and now could continue to extend their control. The peasants' defeat did not change a major process in forest history; rather, it made this peculiar path irreversible.[1]

In 1832 the ruins of the Hambach Castle became a center of national interest when thirty thousand people came together and claimed a united Germany, more political rights, and freedom of press and opinion. This demonstration represented a climax in the liberal movement in nineteenth-century Germany. The castle ruins have become a national symbol for the suppressed democratic tradition in Germany.[2] During the Hambach Celebration and some days after, many attacks occurred on the Bavarian foresters and the local administrations. The rural population claimed a different forestry, considering their problems of providing themselves with wood and other forest resources. The forests had changed a lot in the past decades: where one would once have found clearings and fields inside the forests, one could now see upcoming young trees, mostly conifers like pines and spruces. If one were close to the next village, one would not find dead leaves or little twigs on the ground—rather, the forest would look as if it had recently been swept.[3] One might see some families with rakes on their shoulders, carrying rags full of leaves on their heads. The collected litter served for the stabled cattle and later as a fertilizer for the fields and vineyards. And one probably would be stopped by a man in a green uniform, halting any "visitors" and asking them to retrace their steps.

These highlights in the history of the famous wooded hill from 1500 on illustrate the extent to which forests have changed throughout the centuries. Leading up to the twentieth century, the agroforestal system of preindustrial times has completely disappeared, the multifunctional woodlands have been transformed by a professional forestry that had to produce a maximum of timber. In the past two decades of the twentieth century, the prices of wood and timber have fallen, which have made a new definition of the tasks of forestry possible. The question of whether the periodization of environmental history—here the history of forests—coincides with the periodization of history in general has to be treated with care. The highlights show that this question cannot be answered in general. On the one hand, the Peasant War was a turning point in German forest history; on the other hand, the liberal movement of the mid-nineteenth century and the Hambach Celebration was not such a turning point, although the rulers felt threatened as much then as in 1525.[4] Forest history, as a part of environmental history, needs its own periodization, which can coincide (and often does) with common chronological schemes. This is only the case for turning points that concern societies in their entirety, however. Successful revolutions or global wars indisputably belong to that category.

The Problem of Turning Points in Forest History

Defining the turning points in forest history is a difficult, if not impossible, task. Four major problems have to be solved before one can find a scheme of periodization in forest history that is valid for many Western societies from 1500 to today. Trees grow very slowly, so one must ask how to identify turning points in their history, if their life spans far surpass the life expectancy of humans? Changes in forestry often show their effects more than a century later, which makes it difficult to precisely define a turning point. On the one hand, an oak that had been planted in the days of the Hambach Celebration may be cut down thirty or forty years from now. On the other hand, an entire forest could quickly be cut down. The destruction of forests by historic events like wars can also mark turning points. Insisting on the idea of "turning points" in forest history also means stressing the aspect of destruction—an aspect that should no longer be the one-and-only leading concept of the discipline.[5] Because trees grow so slowly, constructive changes in forestry (like afforestation) need a lot of time, so it may be more accurate to speak of phases of transformation of a forestry system or phases of transition to a new type of forest.

The development of forests in Western countries varies significantly in natural conditions. How does one bring together the development of such different forests as the Mediterranean Garrigue (scrubland), the forests in the Austrian Alps, in the Black Forest, the Baltic Pine forests, the deciduous woods of low mountain ranges, and the forests at the Polar Circle? The climate, soil, fertility, humidity, and slope differ a lot within each of these forests, and so does the human impact. There is no forestry system that could be applied under all of these various conditions. Forestry inevitably has to be adapted to these different natural conditions. This makes it uncertain whether a chronological scheme can be found that is valid for all Western countries. Like every other environmental historian, one has to integrate all the different developments in space and the simultaneity of nonsimultaneous behavior. For example, in the mid-nineteenth century, when German and French foresters reafforested many clearings and forbade cattle pasture in the forests, the famous routes of transhumance on the Spanish Meseta was still practiced and prevented any major afforestation in Castile.[6] The basic issue at stake is the possibility of defining a periodization in forest history that is valid beyond national boundaries.

Five Dimensions of Change and Continuity

Major trends in the political dimension of forestry include the increasing influence and power of the rulers (and later the state) who established a severe forest administration.[7] This broke up local traditions to implement a new forestry—first on a regional and later on a national and imperial level. The strength of the state's impact was slowed down but rarely stopped by the resistance of other forest owners or users. Finally, they all accepted the state regulations. The history of forests is strictly related to the history of statehood. In the long run the territorial rulers and later the nation-states became the most important agents in forestry. With the printing technology of the sixteenth century, hundreds of *Forstordnungen* (forest orders) were published in Central Europe claiming deforestation and overexploitation of the forests. The basic issue of these claims was not really the endangerment of the forests, but the establishment of power and control over the forests.[8]

Even the smallest states enacted different forest laws, but no law was able to prevent local populations from using the forests. In some regions the fines for illegal use of the forests were much higher than the financial output of the timber trade. However, this control was ineffective as the number of foresters and forest guards was too little to prevent any danger. This changed with the administrative reforms in all Napoleonic and post-Napoleonic territories. A higher number of foresters made a better organized administration of forests and a severe control of its resources possible. The reforms had been necessary because of the high financial debt from the revolutionary and Napoleonic wars.[9] A major motive to establish a powerful forestry was the financial gain the rulers could obtain by selling the timber, for example, to the Dutch.[10] And many states had to provide fuel to their state-owned ironworks too. In principle, the state forestry was a kind of state-owned enterprise, like mining or railway companies. During the nineteenth and twentieth centuries most forest administrations received high profits from selling wood and timber. Compared with these gains, the costs of reafforestation were very low.

For states the formation of professional forest managers was a first priority, so they founded forestry schools.[11] Contrary to the dominant liberal ideal, the forest administration also extended its control to the woodlands communities and the commons, trying to force a divide among members.[12] A measure of independence was left to the community councils, as they could decide what to do with the results of the cutting. Yet their influence was limited, because the forest district and the extent of any cut had to be fixed by a professional (that was usually a state) forester. The aim of this state-controlled forestry was to ensure the sustainability of wood production. It broke up local traditions and established forestry first on a regional then on a national level.[13] The political dimension of forest history is marked by a continuous strengthening of the state's power. At the end of the twentieth century NGOs like Greenpeace, Robin Wood, and others organized across national boundaries and gained an increasing influence on forest policy. Some of their former activists now occupy key positions in state administration in Berlin, but also increasingly in transnational organizations in Brussels, New York, Rome, or Nairobi. Forest politics is no longer merely a national matter.

The path of the economic development of forests leads from the multifunctional use of the woodlands (which made property less important than use rights) to state-owned enterprises (which were concentrated merely on timber production), from a common use of forest resources to commercialized goods, from a time when forests delivered many key resources for the economy up to their substitution, from local to global markets, from economic periphery to global networks. In his famous book *The Modern Capitalism* the economic historian Werner Sombart has described the era before industrialization as the "Wooden Ages." He emphasized that wood and timber were the material basis of preindustrial production and everyday life.[14] Wood was needed for many reasons: as fuel, as raw material for the crafts, as timber for ship and house construction, to burn charcoal for iron founders, to cook potash, and to produce glass and soap; tanbark was needed in leather production; pitch and tar for different purposes. The Wooden Ages reached their climax with the historic debate about the shortage of wood. But in this debate the participants only focused on wood and timber production. This made it possible that a century later, economic, rural, and forest historians could also neglect the high importance of the woodlands for the peasant's economy for pasture, litter-collecting, or as additional land. Historical environmental research unanimously shows that up to the nineteenth century, forests must be considered an agroforestal system, that was heavily opposed by the foresters.[15] Because of this debate on "wood famine" and the financial interests of the forest owners, the agrarian users were increasingly banned from the forests. The Wooden Ages ended with the energy revolution, when firewood and charcoal were substituted with coal, potash with soda (sodium oxide), and litter by guano and artificial fertilizers.[16] The forests thus lost their key function for the economy.

The era of commercialization started in the mid-eighteenth century, when financial interests became more relevant for the forest owners. Timber and firewood became goods that were sold to the highest bidder. It might be argued that the economic transformation began earlier, with the famous Dutch timber trade downstream. The rivers in northern and northwestern Europe were essentially part of international timber trade, which substantially affected the forests. But this trade only touched those forest regions where the required timber for construction could be found, which were close to floatable rivers that ran toward the Netherlands. In the nineteenth century an increasing portion of firewood and tanbark was sold and became goods that were sold to a local and regional market too. Commercialization reached even the smallest villages. With the construction of railways, even the remote mountain forests were integrated in a wood market that no longer depended on the river systems. Mass import of timber and wood from the Baltic also affected wood prices in Central Europe. From the 1860s on, even countries like Germany, which compared to other industrialized countries had many forests, depended on wood and timber import.[17] The imperialism at the end of nineteenth century and the falling costs of transport by ship and by train favored the formation of an increasingly international and global wood market. Formerly peripheral forests were gradually integrated into global economic development.[18]

The social dimension of forest history is marked by population growth, by increasing social exclusion, by migration, and by the establishment of an "expertocracy." Growing population is one of the key issues of environmental history and thus for forest history. With an increasing population, the demand for firewood and timber rose, as did the number of livestock and the need for pasture. In medieval times the rural population was relatively free to use the forests, to cut firewood or to put out their cattle for pasture. With the rise in population in the eighteenth century, however, these use rights had to be limited to protect the forests from destruction by overexploitation. Some of the uses were forbidden, others were paid off by the owners or the forests were split up among the owners.[19]

In the nineteenth century a large part of the common woodlands were divided too. Access to forest resources once organized by traditional use rights for a community now was supplied by a market. This made it difficult for the poor to gain their livelihood. Without money they frequently had no alternative but to supply themselves illegally with fuel or litter.[20] At the end of the nineteenth century the big migrations overseas and to the industrializing cities made this problem less severe. But migration from the country was much more important in the twentieth century, when farming became more industrialized and many peasants gave up their small and unproductive farms, searching instead for work in towns. Nowadays, forests have again become an open space to which everyone has access—but not for economic purposes.

Since the end of the eighteenth century, the control of the forests has passed into the hands of a new profession, the foresters. The science of forestry was developed and taught in forestry academies like Tharand (Saxonia), Eberswalde (Prussia), or Nancy (France). The foresters formed in these institutions monopolized forest management and thus formed an "expertocracy."[21] When speaking of the cultural dimension of forests, I do not mean the way of thinking and presenting forests in arts and literature. More important is the perspective of historical ethnology. We should examine the way people who lived in forests or who used them actually thought about the forests and legitimated their behavior. There have always been different ideas as to what a woodland is, so this cultural field can better be described by poles than by clear steps of development. These poles might be stability and growth, the striving for livelihood and for profit, personal or common interest, and, of course, sustainability and destruction.

The traditional preindustrial use of woodlands was determined by the everyday need of fuel and fodder. The majority of the rural population did not search to maximize their profit, but to gain their livelihood. Timber merchants in the seventeenth and eighteenth centuries were the avant-garde of a new type of men, who made timber an export commodity. They introduced the idea and the possibility of profiting from forests. The rulers, whose main interest in forests had been hunting, now changed their focus to financing their castles and courts. Hunting was still important to demonstrate power, but from the second half of the eighteenth century on, "profit" became the magical word of forestry. The foresters did not see any contradictions with the ideas of sustainability they propagated.

"Sustainability" at this time did not mean the conservation of a given ecosystem, but the continuance of wood and timber production to make a sustainable profit. Sustainability was an economic not an ecological value. This changed in final decades of the twentieth century, however, when sustainable development was redefined by environmentalists and the whole system of economic forestry came under pressure. Nowadays "ecology" and "economy" are considered no longer to exclude each other, but at least in principle to be compatible concepts.[22] This occurrence has changed the way foresters view themselves. They no longer set the agenda of forest politics and controlled the discourse on the tasks of forestry; rather, they had to consider the ideas of the ecological movement. To describe it according to Gramsci's concept of cultural hegemony, these new ideas at first seemed to be completely absurd. When they were disseminated, they were heavily fought (by the foresters), but finally these fresh thoughts were integrated in a somewhat transformed system.

The ecological dimension of forest history is marked by an interaction between humans and the rest of nature. The most important developments are the shift from heterogeneous woodlands to homogeneous forests, from biodiversity to monoculture, from extensive uses to an intensive use of the forests, from overexploitation and destruction to sustainability. The forests of the sixteenth century (described at the outset of this chapter) have completely disappeared in modern times. Once the woodlands had been an area utilized by people for many different purposes: wood collecting, timber cutting, charcoal burning, farming, collecting leaves, grazing cattle, litter collecting, blueberry picking, mushrooming, beekeeping, and many other uses. The various combinations of these uses in changing intensities made woodlands heterogeneous. The appearance of a forest differed considerably from one place to another. In some cases these various uses destroyed a once existing forest by overexploitation. Because trees grow so slowly, it is difficult to observe changes in the forests; change takes a long time to become visible.

The combination of several forms of uses on a limited area destroyed some forests, like a combination of intensive cuttings for charcoal-burning followed by farming and cattle grazing, which prevented the sprouting of young trees. Such an area was not ultimately recovered by trees, but instead exposed to desertification or erosion. The claims in the forest laws of the population's destruction of forests were more than just an argument to extend the rulers' power and to fight competing interests.[23] But speaking of destruction and degradation introduces a judgment of values into the debate.[24] Is a forest that changed its character from an oak forest to a pine forest degraded? I confine the discussion here to describing those changes that occurred as a result of human interaction with the natural environment.

Obviously, overuse threatened many forests with diminished productivity or even ruin. State intervention was often necessary; the sustainability of many woodlands was endangered. Paradoxical as it may seem, in an ecological view the effect of the new professional forestry was far from being purely positive. In most forest stands today only one or two different tree species is found, all of the same age and size; there are very few clearings and glades. From the beginning of the twentieth century on, most central European forests could be characterized as uniform and standardized. Why? In the early nineteenth century, starting in Germany, it was essential for the forest administration to survey the growing of trees and to plan which trees should grow in which part of the forest and when to cut them. The less heterogeneous the forest was, the easier to plan its growth and to guarantee the sustainability of the wood production. The exclusion of agrarian uses became visible in the landscape too. Where once a distinction between pastures, meadows, and woods was difficult to make, at this time a clear and sharp line between farmland and forest could be observed. The new type of forestry preferred timber forests and therefore transformed most coppices into high forests. Trees had to grow for a much longer time before they were cut.[25]

The reduction of the forests to one main function was also a reduction of biodiversity as the former multifunctional woodlands had had many more different species than the "industrialized" forests at the end of the nineteenth century. By eliminating the undesired species and trees of a different age, those forests became more standardized.[26] A shift from deciduous to coniferous forests and many monocultures as the most extreme form of these monofunctional forests have been other consequences of that forestry. No wonder that some storms in the 1990s heavily damaged those forests. They are also less resistant to acid rain and diseases. But the negative effects of industrialization (like air, water, and soil pollution) also befall forests that show a higher biodiversity. Negative effects on forests that were endogenous were relatively easy to combat with strict forestry, but today there are increasingly exogenous factors that cross national boundaries. As an effect of the storms and of a globalized market, the prices for wood and timber fell and made forest administrations for the first time in their history enterprises with deficits. This development ushered in a discussion about the task of forestry, transforming it in a way that was less orientated toward pure economic values.

From Local to Transnational Turning Points

Why do we need turning points in forest history? The most important task of turning points is a didactic one. To teach forest history, we need to set some benchmarks, the turning points that should help orient time. These dates symbolically represent an important development that has started with an event or could be symbolized by it. It is not difficult, for example, to define turning points in the forest history of France or Germany. A turning point in French forest history was the nationwide organization of forestry by the French minister of finance, Jean-Baptiste Colbert, in 1669. Another turning point was in 1789, when the nobility and the church were expropriated by the revolutionary Assemblée Nationale. These two events were key issues for French forest history, but had no effect on Germany. Germany's Peasant War of 1525 did not influence France, nor did the Thirty Years War in the seventeenth century, which caused a decline of population in many German regions and a recovering of soils with trees as well.

It seems as though it is impossible to even identify common turning points for two neighboring states. But this view does not hold up, because from the nineteenth century on, similar developments can be observed in both France and Germany. Both countries were industrializing and became more dependent on the import of wood and timber for mining and industrial purposes. In both countries firewood was increasingly substituted by coal, the agrarian users were banned from the forests, and reafforestation movements also appeared. In the early twentieth century, World War I weighed heavily on both French and German forests, and after World War II a lot of timber was cut for reconstruction. In the 1960s and 1970s a second and bigger wave of rural exodus started, and more people left the countryside. From the energy revolution in the mid-nineteenth century on, French and German forest history have more or less similar turning points.

This observation, which can be verified when comparing other neighboring states, leads to the following hypothesis: It can be maintained that turning points in early modern times can only be defined on a local or regional—in cases of early national states also on a national—level but not across boundaries. When the Western states began to enter the field of global politics and economy—in times of high imperialism—the political, economic, and social developments in Western countries were linked so closely to one another that it is possible to identify common turning points not only in forest history. This linking-up of the Western countries from the 1870s on fits another observation very well. The impetus for big changes in forestry was rarely an endogenous (in most cases it was an exogenous) change that caused a reaction. Transnational developments became more important than the decisions of national forestry administrations. Of course that did not put an end to all visible differences in forestry. These differences still exist, but on a far smaller scale.

So what are the major turning points in the field of forest history? The first transnational turning point in forest history can be seen in the Napoleonic era, when nearly all European countries more or less adopted the French administrative system and the Napoleonic Code. Even countries that developed their own administration copied many structures from those systems, at least the specialization of administrations, which meant an independent and specialized forest administration. There was a financial necessity to realize those new and more efficient structures. Forestry was now a science and a profession; its main task was to produce more timber for increasing markets, so the agrarian users were more and more excluded.

A second turning point was marked by the construction of railways in all Western countries. This made it possible to substitute firewood by coal. But this was also a revolution in transportation that integrated the forests into an international economic market, independent of nature (mountains, river systems). This turning point was less important for states like the Netherlands or England that were much more orientated toward transport by sea and less limited by natural obstacles.

The third turning point has often been characterized as high imperialism and the beginning of globalization. The economic development and demand for timber and other forest resources of the different nation-states was so closely linked that all the economic crises, like the Great Depression of 1873, took place at the same time. The rivalry of the European nation-states finally led to World War I. The two world wars in the twentieth century were of course important turning points for history in general, but besides the destruction of warfare and the needs for reconstruction material, there was little effect on the development of forestry. Much more important was the fact that in the 1950s the prices for petrol sank so much that labor became much more expensive than energy (the "1950s syndrome").[27] Another big wave of industrialization and automatization increased industrial production, which produced a heavy effect on the forests: pollution.

The emergence of the environmental movement from the 1970s on is another turning point. The movement had an increasing influence on politics in general and on forest politics specifically. Forestry was no longer the monopoly of foresters, but a field that was disputed by environmentalists. At the turn of the millennium, forestry had integrated environmentalism on a large scale and did sometimes even change the name of their faculties.

5

The Nation-State

THOMAS LEKAN

◯ At first glance, analyzing turning points in the environmental history of the nation-state might appear to be a departure from a central goal of the field: to overturn politics as the basis of historical inquiry. During the founding years of environmental history in the 1970s, esteemed environmental historian Donald Worster has noted, "historians lost some of their confidence that the past had been so thoroughly controlled or summed up by a few great men acting in positions of national power."[1] They began to excavate the "hidden layers" of class, gender, and race as the true agents of history until they hit the most important substrate of all: the natural environment. The field's central themes, as Worster explains—be they the Dust Bowl of the 1930s, Tuareg nomads in the African Sahel, the great whales of the oceans, or global patterns of climate change—defied national boundaries. Worster proposed that such themes demanded an expanded Marxist analysis that integrated the natural environment into its base, resulting in a tripartite structure: natural ecosystems; modes of production; and values, ideas, and perceptions. Although he ascribed value to all three levels of analysis, Worster, like many environmental historians of his generation, privileged the first and second levels of analysis as the foundation of environmental history. "Ideas should not be left floating in some empyrean realm, free from the dust and sweat of the material world," he wrote. "They should be studied in their relations with . . . modes of subsistence . . . the historian must understand that mental culture does not spring up all on its own."[2]

Noticeably absent from Worster's concept of environmental history is any mention of the nation-state as an instrument of ecological change. His underlying assumption equates the nation-state with a rarefied circle of dead white men,

55

diplomatic maneuvers, military campaigns, or sweeping assumptions about the "national mind." But in the years since the publication of his essay in 1988, historians have reinvigorated political history with models of the nation-state as an "imagined community" characterized by historical contingency, spatial disruption, and social engineering, rather than the teleological pinnacle of historical development. As the esteemed Harvard historian Charles Maier has noted, the emergence, ascendancy, and subsequent crisis of what he labels "territoriality" provides the most significant historical development of the past two hundred years. Maier describes territoriality as "the properties, including power, provided by the control of bordered political space, which until recently at least created the framework for national and often ethnic identity."[3] In his view such fixed territoriality emerged around 1860, reached its heyday in the early twentieth century, and underwent a period of crisis around 1980.

This scholarly emphasis on the invented nation reflects the post–Cold War era of globalization, when time-honored political structures seem to have evaporated overnight. Environmental historians have an important stake in this new national historiography, for the era of fixed territoriality had a profound impact on societies' relationship to the natural world. Indeed, many central narratives of environmental history—struggles over property rights and access to natural resources, metropolitan elites' promotion of technological development in the provinces and colonies, the identification and control of disease pathogens, and the emergence of nature conservation and environmental protection movements—are scarcely imaginable without a nuanced understanding of the nation-state and its representatives.

To explore ways that environmental historians might demarcate the turning points of nation-state environmental history, this chapter uses anthropologist James Scott's highly influential periodization of state building in his *Seeing Like a State: How Certain Schemes to Improve the Human Condition Have Failed*.[4] As Scott has demonstrated, the rise of the modern state in the eighteenth century empowered authorities who tried to fix territorial boundaries, facilitate taxation, manage conscription, and boost economic growth through an "ongoing project of legibility" that simplified human and ecological connections.[5] He shows how centralized planning and high modernism often have led to ecological and human disaster through a series of case studies: German forestry; Le Corbusier–inspired city planning in Brasilia and Chandigarh, India; Soviet agricultural collectivization; Tanzanian "villagization"; and high-yield mechanized agriculture. In the tradition of such sociologists as Max Weber and Michel Foucault, Scott sees these disasters as the result of modernity's particular configuration of knowledge, institutions, and power.

In this view even the earliest "disinterested" acts of mundane statecraft—such as recording harvests, gathering census data, conducting cadastral surveys, or identifying natural resources—presupposed radically remaking space and society according to a visually coherent, rationalized grid. Scott believes that this logic of rationalization transcended political ideology, especially in the twentieth century, and applied equally to the "private" sphere of large corporations and to massive public bureaucracies, socialism as well as capitalism. In his words, "large-scale capitalism is just as much an agency of homogenization, uniformity, grids, and heroic simplification as the state is, with the difference being that, for capitalists, simplifications must pay."[6]

Scott's analysis thus scrutinizes the social and spatial strategies of territoriality in highly aesthetic terms. It charts a culturally specific way of demarcating, abstracting, organizing, and disciplining society and the natural environment that dominated the nation-state's "spatially anchored structures for politics and economics."[7] Such a perspective disrupts the neat division of environmental history into biological, economic, political, and cultural realms, not by dissolving the material world into "discourse" but by showing how state institutions narrowed and simplified the range of what was worth knowing about the natural world to a few instrumental categories: "forests" became "timber," "meadows" became "pastures," "nature" in general became "natural resources."[8] *Seeing Like a State* juxtaposes such a top-down, utilitarian optic with *mētis*, an informal, local, on-the-ground lens that underwrites complex social activity (and, environmental historians might add, practical sustainability).

While Scott developed his model as a contribution to historical sociology, numerous environmental historians have found the work useful for thinking about human-nature interactions, especially the conflict between imperial states and various subaltern groups over access to natural resources.[9] Scott's model also leaves plenty of room for environmental historians to discuss the ecological blind spots of statist myopia with accounts of nature's agency at transnational and ecosystemic scales. Whether they are discussing the vulnerability of timber plantations to insect infestation or the role of customary privileges in maintaining the health of woodlands, environmental historians have shown again and again how "seeing like a state" has left an array of natural processes, plant and animal species, local modes of production, and indigenous botanical knowledge out of its purview, often to its own peril.[10] It was also coercive; as these examples show, "the human domination of nature also bore eloquent testimony to the nature of human domination."[11]

This chapter considers several examples from the environmental-historical literature on forests, hydrological engineering, and nature conservation to demonstrate the value as well as the limitations of Scott's synthetic framework for designating turning points in different national contexts and at different scales of analysis. The chronology of nation-state environmental history developed in this essay, loosely based on Scott's analysis, is a familiar one. It has affinities with environmental historian Carolyn Merchant's concept of "ecological revolutions," in which Western societies developed radically new sets of relations with natural systems at all three levels of environmental-historical analysis: the ecological, the economic, and the cultural.[12] If, as Charles Maier has noted, periodization is an "attempt to interpret more than to explain," then scholars can use these chronological breaks to frame a "grand narrative" of environmental transformation in the modern age. The narrative begins in the seventeenth century with new utilitarian and scientific frameworks for understanding social relationships with the natural world and continues into the eighteenth century with an emerging profit-driven, resource-extractive, and fossil-fuel dependent economic order. It follows the spread of this economic order in the nineteenth century across the globe through colonialist and imperialist expansion, and ultimately charts the widespread realization in the twentieth century that these processes had impoverished species and habitat diversity, polluted the land, air, and water, and even changed the earth's climate. These changes coincided with the rise of the modern nation-state, yet few historians have analyzed how and why this political structure has shaped (and was in turn shaped by) these new patterns of environmental interaction.

Putting the state back in as an agent of environmental transformation enables scholars to assess the ecological foundations of modern territoriality in a fresh light: as a risky, highly vulnerable, and often catastrophic attempt to engineer society and control nature according to simplified, rationalized, mechanistic models. For the purposes of environmental history, one should see the boundaries between Scott's phases as fluid "contact zones" rather than as concrete turning points.[13] Most striking, however, is how few state officials and policy experts learned to see *beyond* the state, even in the more ecologically conscious age after 1970. Developing nations have mimicked early-twentieth-century development schemes, and supposedly "green" infrastructure projects have often displayed a callous disregard for local traditions and needs.

The Rise of the Modern European State, ca. 1700–1850

The rise of the modern state and its utilitarian ethos dominates Scott's first phase of analysis. His major case study, German forestry, is familiar to most environmental historians and reveals in highly aesthetic terms the statist desire to bring a complex, unwieldy entity—that is, the natural forest—under the simplifying gaze of forest scientists and planners. It remains a matter of debate whether a real "timber shortage" prompted the first attempts at forestry management in Prussia and Saxony between 1765 and 1800.[14] But state interest in ensuring trees for military vessels, fuel for its subjects, and saleable timber took numerous parcels of forested land out of the commons and into state hands to ensure a predictable and steady supply of lumber. Such cameralist utilitarianism elevated revenue generation above all local peasant needs. Statist myopia reduced trees' myriad uses to board feet of commercial lumber or cords of firewood for sale, bracketing out the foliage for fodder and thatch, the fruits that people or domestic animals might eat, the twigs and branches that might serve as bedding or fencing, or the barks and twigs that had medicinal purposes.

Not seeing the ecologically robust and multiuse forest through its commercially valuable trees soon led foresters to replace the mixed-use, uneven-aged forest altogether in favor of conifer farms, a process that paralleled the broader enclosure of customary open agricultural fields. These changes shifted power from "those who knew the land to those who knew the law," but not without a fight.[15] Advocates of a "moral economy" in the countryside refused to give up easily on the forest commons for timber, animal foraging, and gathering, and turned instead to poaching, insurrection, and even arson to thwart state efforts. Nature also responded to the elimination of species diversity with an invasion of insects, rodents, and diseases, crippling storms that toppled same-age and same-species stands, and a thinning and depletion of soil nutrients. The initial success of German scientific forestry nonetheless led to its export to France, England, the United States, and European colonies, making this story a global one. However, indigenous resistance made the imposition of colonial forestry regimes a highly contested and precarious undertaking, as environmental historians of India and Africa have demonstrated.[16] One can find echoes of state control, visual simplification, and local resistance in other case studies from absolutist and parliamentary systems: the monumentalism of Baroque city planning, the fixed property regimes that justified the dispossession of aboriginal peoples in the Americas and Australia, Baden engineer Johann Gottfried Tulla's "correction" of the Rhine River, or the attempt to tame tidal waters and discipline enslaved Africans in the "hydraulic machines" of Carolina and Georgia rice plantations.[17]

Scott's focus on state simplification leaves out a powerful intellectual trend that challenged the utilitarianism of the cameralist state and its massive transformation of the old regime landscape by glorifying nature's sublime and primeval energies: Romanticism. German painter Caspar David Friedrich's *Chausseur in the Forest* (1814), which depicts a lone soldier returning to his sylvan home after the Wars of Liberation against Napoleon, was one of dozens of works in the German-speaking lands that combined forest Romanticism, nascent German nationalism, and a rejection of French classicism. Around this same time Rhine Romanticism emerged, first as a literary and artistic phenomenon glorifying sublime nature, later as a tourist brand marketing the river to British visitors. After the Napoleonic Wars, British tourists who had been prevented from visiting the continent flocked to the Rhine in droves to follow in the footsteps of such luminaries as Anne Radcliffe, Mary Shelley, and Lord Byron, by viewing for themselves the gothic ruins, billowing clouds, and twisted rocks of the Slate Mountains that had inspired these writers' florid prose and poetry. Such romanticization went hand in hand with the river's canalization and industrialization. In 1822, Heinrich Heine published *Die Lorelei*, a poem about a mythical siren along the Rhine who lured unwary sailors to their death in the reefs below the surface. This was just five years after Tulla had begun the "rectifications" of the river that shortened its length, eliminated braids and islands, and cut a new, straighter bed that made the river resemble "an industrial faucet more than a natural river."[18]

Intellectual historians such as Clarence Glacken and Roderick Nash have identified the period around 1800 as a turning point in humankind's ethical stance toward nature, as the Romantic cult of the sublime and the picturesque led Western culture to appreciate the intrinsic value of nature for the first time. This cult, which saw a divine hand in the untrammeled wilderness, eventually led to the creation of the world's first national park at Yellowstone in the western United States.[19] Yet Rhine Romanticism did little to stop the fundamental forces of hydraulic transformation; if anything, the development of modern tourism depended on and even accelerated the transformation of the riparian environment. Beginning in the 1820s, tourists enjoyed their sublime and picturesque vistas of such famous landmarks as the Drachenfels or the Lorelei from the deck of passenger steamships that shortened the fourteen-hour trip from Rotterdam to Cologne to just over five hours. In 1827 the Prussian Steamboat Company (predecessor to the now ubiquitous Köln-Düsseldorfer) started serving the scenic Middle Rhine Gorge from Cologne to Mainz. In its inaugural year the company's steamers carried eighteen thousand passengers; by 1913 that number had surged to 1.9 million.[20] They flocked to the sinuous and hilly Middle Rhine because the Upper Rhine, Tulla's Rhine, was deemed "dull" by guidebooks. Yet even on the Middle Rhine Gorge, the steamships, like industrial barges, depended on a channelized river for safe navigation, which enabled visitors to experience "nature" as an unfolding panorama unencumbered by reefs, braids, gravel banks, and seasonal variations in flow. The cultural construction of the "wild" Rhine for tourism was thus dependent on its "taming" by engineers; the Romantic gaze did not pose a fundamental challenge to the utilitarian statist one.

Utopian or Authoritarian High Modernism, 1850–1970

In Scott's second phase of development, the growth in confidence about scientific and technical knowledge, the expansion of production and material well-being, and a faith in increasing control over nature and society spawned by Enlightenment thought and statecraft combined to shift the state's rationalizing tendencies from description and extraction of wealth and resources to prescriptive social and environmental engineering. As Scott has noted, eighteenth-century state officials had "mapped more of their kingdom's population, land tenures, production, and trade than their predecessors . . . [and] . . . had become increasingly efficient in pumping revenue, grain, and conscripts from the countryside."[21] Yet the claim to absolutist rule belied a lack of military might, policing powers, administrative capacity, and knowledge to undertake "more intrusive experiments in social engineering." This situation changed after 1850, when "the existing social order, which had been more or less taken by earlier states as a given, reproducing itself under the watchful eye of the state, was for the first time the subject of active management."[22] Only in the mid-nineteenth century in Europe and in the twentieth century elsewhere were the conditions present for a breathtaking, muscle-bound faith in social and environmental transformation that Scott has labeled "high modernism," a periodization that coincides exactly with Maier's rise of "territoriality" as the dominant principle of political organization.

The avant-garde among state officials, engineers, technocrats, architects, scientists, and revolutionaries from both the right and the left who envisioned these sweeping changes tended to view their task as above political rancor and emancipatory in its potential. They promised freedom from scarcity, want, and the arbitrariness of natural calamity, as well as liberation from the dark side of human nature: religion, superstition, and irrationality. But the apex of high modernist ideology occurred in the twentieth century, when industrial warfare and worldwide depression required total mobilization of belligerents' social, economic, and natural resources. In Britain, for example, the loss of one-half of the country's productive forest land during the First World War led to rapid passage of its first Forestry Management Act in 1919. In North America the 1936 propaganda film *The Plow That Broke the Plains* blamed the rising wheat prices during the conflict for the frenzied development of the ecologically fragile High Plains, paving the way for the Dust Bowl disaster of the early to mid-1930s and massive federal intervention through the Soil Conservation Service and the Rural Resettlement Association.[23]

In liberal, parliamentary societies such as Great Britain or republics like the United States, ideas of the private sphere, free market, and representative institutions could prevent high modernism from running amok, even though the logic of economic progress usually prevailed over environmental concerns. But in authoritarian nation-states severely weakened by wars, revolution, civic unrest, or colonial domination, high modernism became lethal. As Scott has put it: "The ideology of high modernism provides, as it were, the desire; the modern state provides the means of acting on that desire; and the incapacitated civil society provides the leveled terrain on which to build (dis)utopias."[24]

Scott's most dramatic examples of authoritarian high modernism in motion occurred in rural resettlement and agricultural improvement strategies dictated by centralized elites—the forced Soviet collectivization of the early 1930s and the compulsory Tanzanian "villagization" campaign of 1973 to 1976. In the Soviet Union "a combination of defeat in war, economic collapse, and a revolution had provided the closest thing to a bulldozed site that a state builder ever gets," wrote Scott.[25] In Tanzania, however, a more benign and weak postcolonial authority never realized its plan fully due to an unwillingness to employ Stalinist methods and local peasants' tactical advantages, which included flight, smuggling, and foot dragging. According to Scott, both schemes failed because they neglected or actively suppressed the informal, noncodified practices and improvisations that enabled the rural population to deal with the vagaries of climate, disease, fertility, and economy for centuries.

Environmental historians will recognize the high modernist era as a period of "brute force technologies"—such as large dams, massive factories, mechanized farms, expansive railways and highways, and centralized urban planning—that sought permanent mastery over capricious natural systems.[26] Such technologies, according to the historian of technology Paul Josephson, "are not merely big technologies; they include the scientific institutes, financial organizations, engineering and construction firms, and government bureaucracies drawn together to transform nature in the interests of citizens."[27] In Josephson's view such technologies play a powerful but neglected role in modern environmental degradation, which scholars often attribute to population pressures or the rise of capitalism while neglecting the self-perpetuating hubris of technological optimism as an independent factor. The short-term efficiency of these technologies, moreover, makes them appealing across political ideologies; indeed, Josephson claims that the "Stalinist plan for nature transformation" has been under way in all industrialized nations, especially in the United States, throughout the twentieth century. Until recently, the social and environmental costs of such know-how could be rationalized because the technologies produced real gains in employment and domestic well-being for urbanites and placed great power in the hands of technicians and their sweeping vision of modernization.

No technology better exemplifies the state's high modernist impulses than hydrological engineering and dam-building; indeed, the modern hydroelectric dam "serves as . . . [the] . . . exemplar of human aspiration and folly. With its scale and aura of permanence, it drowns out human agency both literally and figuratively."[28] The dam-building craze in Europe and North America began around 1900, reached its heyday in New Deal America and the Soviet Union of the 1930s, and (at least in the West) began to decline after the 1970s. While beavers and some humans have been damming rivers as long as either species has existed, the scope and scale of dam building in the twentieth century was unprecedented. When the World Commission on Dams surveyed the globe around the year 2000, it counted more than forty-five thousand large dams; in the heyday of dam building in the 1970s, two or three dams were commissioned on average each day.[29] Proponents of dams extolled the seemingly endless water and cheap energy that dams would provide, naively unaware of the ecologically destructive capacity of dams that came to light in the 1970s.[30]

Although the first hydroelectric power station began operation on the Fox River near Appleton, Wisconsin, in 1882, it was the Boulder Dam (later renamed the Hoover Dam), built in the 1930s, that ushered in the twentieth-century's large-scale dam-building craze. Under the aegis of the U.S. Bureau of Reclamation, the federal government set out with missionary zeal to make the high plains and deserts bloom. In his landmark study *Rivers of Empire* (1985), American environmental historian Donald Worster has argued that the Hoover Dam and similar large-scale impoundments were essential to the demographic and economic growth of the arid U.S. West.[31] In line with Scott's analysis of authoritarian high modernism, Worster showed that the need for water not only irrevocably changed many of the landscapes and ecosystems of the West but also led to a radical redistribution of power in favor of technocratic elites.

Following in the footsteps of Karl Wittfogel, a German-American scholar of Chinese civilization, Worster argued that the hydraulic apparatus in the West created a vast bureaucratic apparatus operating beyond the democratic process. The American West, quintessential landscape of freedom in the American imagination, was in reality a despotic regime designed to control nature and people. But the hydraulic regime extended farther eastward; the river-basin management, multipurpose dams, and modernization philosophy of the Tennessee Valley Authority (TVA) also inspired imitators throughout the world, including similar networks along the Columbia River in the Pacific Northwest, the Volga and Dnieper rivers in the Soviet Union, and the Ganges River in India. By the 1990s almost two-thirds of the globe's streamflow passed over or through dams.[32]

While access to water, flood control, electricity generation, and recreation by themselves or in combination were driving factors for dam building, their construction and completion had important symbolic functions, particularly among Cold War competitors and in the recently decolonized countries in the global south. As J. R. McNeill has written: "Governments liked the image they suggested: an energetic, determined state capable of taming rivers for the social good. Dams helped to legitimate governments and popularize leaders, something the United States needed more than ever in the Depression years, and something Stalin, Nehru, Nasser, Nkrumah, and others sought. . . . Ambitious, modernizing states, especially colonial and newly independent ones with legitimacy problems, showed great fondness for dam building. So did Cold War hegemons, keen to display the virtues of their social and political systems. Their political utility helps explain why so many uneconomic and ecologically dubious dams exist."[33]

In newly independent India, for example, irrigation works watered almost fifty million hectares by the 1990s; the hydroelectric potential of these dams led the country's first prime minister, Jawaharlal Nehru, to refer to the concrete behemoths as the "temples of modern India." Nehru underestimated the real economic, social, and environmental costs of the new temples, including displacement of millions of Indian citizens, bloated promises of power generation and durability, siltation caused by river sediments, the destruction of local fisheries, water-logged lands unsuitable for farming, the salinization of irrigated fields, deforestation, and malaria from standing water. Globally, the price has been even higher: between forty million and eighty million people have been displaced by reservoirs, and the benefits of dams have for the most part been inequitably distributed. Large dam projects have also resulted in the loss of forests and wildlife habitats and diminished aquatic diversity. Once these problems came to light in the late 1960s and 1970s, the decline in dam building was as dramatic as its previous surge, especially in North America and Europe.

The worst environmental disaster caused by an irrigation scheme occurred in Soviet Central Asia in the 1950s. Communist central planners had closed off the Syr Dar'ya and the Amu Dar'ya, which carried snowmelt from the high mountains into the Aral Sea, once the world's fourth-largest lake. The goal was to create a vast, irrigated cotton belt; the result was the rapid desiccation of the Aral Sea, which has lost 90 percent of its source waters and two-thirds of its surface areas. In his recent April 2010 visit to the Aral region, United Nations secretary-general Ban Ki-moon voiced his "shock" at this "environmental disaster" and pledged UN support for the leaders of six independent Central Asian republics to save the lake from obliteration.[34] Without such coordination, it will eventually become a "salt pan the size of Ireland, dotted by a few brackish ponds."[35] This hydrological folly destroyed the sea's ability to moderate the local climate by reducing ambient moisture, which shrank the snowmelt that fed the two rivers, spurred aerial salinization that lowered crop yields and destroyed pastures, eliminated the lake's once flourishing fisheries, and killed off half the mammal species and three-quarters of the bird species by 1990. The irrigated cotton itself is of low quality, and much of the irrigated water evaporates or sinks into the ground.

As Scott has noted in his analysis of Soviet development schemes, Communist planners executed such socially dubious and ecologically catastrophic plans because they truly believed that control of external nature would not only serve economic needs, but also emancipate local residents (in this case ethnic Turks) from ignorance, superstition, and backwardness. The goal was thus not merely to rearrange existing patterns of land tenure or economic activity; Communist planners aspired to remake the "nature" of the indigenous inhabitants themselves. The dam would erase inhabitants' connections to "outdated" subsistence modes and help them realize the benefits of socialist planning. If the goal of the absolutist state was to catalog human and natural systems to discern their laws of development, the high modernist state sought to change both in the interests of ever-increasing productivity and rational progress.[36]

Given the state's enormous investment in the symbolic capital of dam building, it is not surprising that two of the twentieth century's first environmental protection campaigns centered on stopping the construction of dams. Between 1906 and 1909 wilderness advocate John Muir and the Sierra Club led the fight to save Hetch Hetchy Valley in Yosemite National Park. In 1904 the German Homeland Protection League tried to prevent damming of the Laufenburg Rapids on the Upper Rhine. In the Hetch Hetchy case, the U.S. secretary of the interior, James R. Garfield, approved in 1908 San Francisco's petition to dam Hetch Hetchy Valley to secure a drinking water supply and source of hydroelectricity for the city. Hetch Hetchy's borders lay entirely within Yosemite National Park; the petition thus tested the government's resolve to leave national parks untouched as wilderness areas. Muir and the Sierra Club launched a nationwide but ultimately unsuccessful public relations campaign to save Hetch Hetchy that spawned the wilderness protection movement.[37] In Germany the Homeland Protection League wrote that the proposed dam would obliterate a highly scenic series of rapids, "one of the most beautiful landscape scenes in Germany, or indeed the world." The league managed to convince such intellectual luminaries as Max Weber and Werner Sombart to support their campaign, arguing that the benefits of hydroelectric power and the lure of jobs at the electric plant did not outweigh the losses to the area's tourist industry. Such advice was ignored when a regional advisory board and the Baden state government approved the dam project in short order and let the rapids be obliterated.[38]

Nature conservationists lost the battle in both cases in part because they relied on aesthetic beauty, rather than on dire predictions of ecological collapse, to press their case for preservation. Throughout the twentieth-century American West, hard-nosed engineers touted the beauty of blue reservoir-lakes set against desert mountain ranges for nature-loving boaters, swimmers, and fishermen to press ahead with dam building. They failed to acknowledge the rapid evaporation of water in such arid climates or the fluctuations in water level that prevented many species from making their home in artificial water bodies. In Germany engineers and preservationists reached a consensus in the 1930s that dams which were "embedded" in the landscape aesthetically actually enhanced, rather than destroyed, the landscape.[39] Here again, recreational interests espousing a love of nature failed to offer any sustained opposition to hydroelectric dams' rapid ecological transformation of the landscape.

While nature conservationists' and landscape preservationists' search for an environmentally sensitive, alternative modernity achieved limited results, they could in certain circumstances appeal to the state as a guarantor of the public good through a "nonmaterial" discourse that called into question the state's instrumentalist orthodoxy. In Germany, for example, preservationists successfully petitioned provincial and national officials to create a nature park along the Rhine at the Siebengebirge by claiming that nature served a therapeutic purpose in an urbanizing society. Outdoor recreation in nature reserves, in this view, offered ordinary citizens, particularly volatile manual workers and unruly young people, a healthy recreational alternative to big-city attractions and guarded the roots of national character.[40] As Scott has acknowledged, the "unremarkable tools of statecraft" are capable of benign reform when a robust civil society is present, as it was in the United States and Germany around 1900. In both the American and German cases, conservationists appealed to the state, as guarantor of the public good, to protect nature as a recreational resource and a font of nationalist sentiment. Indeed, the turn of the century witnessed the growth of a preservationist movement that sought to protect nature, whether as "wilderness" in the United States or as a "cultural landscape" in Europe, in national parks and nature reserves set aside and administered by national and regional authorities. The presence of a mobilized environmentalist citizenry therefore does have the potential to transform the disciplinary state into a guarantor of welfare, freedom, citizenship, and environmental quality.

Even where the state has responded to environmental initiative from below, however, it has not usually guaranteed environmental justice in a broader sense. Although Romantic nationalism had helped to create the world's first large-scale national parks in the United States and Australia, such seemingly benign acts of preservation were predicated on the state's ability to dispossess aboriginal peoples of their lands and to use force to shift existing patterns of Anglo-European rural land use from hunting, lumbering, and ranching to leisure, consumption, and cultural uplift. As the U.S. environmental historian Paul Sutter has put it, putting the state back into the narrative transforms nature conservation's "heroic" limitations on capitalist resource exploitation into a more familiar story of the "eclipse of the commons," with the state clamping down on marginal lands, aboriginal land uses, and customary hunting privileges that had previously escaped its scrutiny.[41] In creating Adirondack State Park, for example, New York's Forest Commission faced rural folk who had responded to the state's encroachment on fishing, hunting, and timber gathering with local acts of trespass, pilfering, and arson guided by a "moral ecology" unintelligible through the state's aggregating and rationalizing lens.[42]

Undergirding American conservationists' desire to create wilderness or forestry preserves with fixed, carefully policed boundaries was a powerful "degradation narrative" about ordinary people and their place in nature.[43] Such conservationists as George Perkins Marsh assumed that nature, when left undisturbed by human beings, attained a state of static harmony. All human impact was therefore deleterious: "Man is everywhere a disturbing agent," wrote Marsh. "Wherever he plants his foot, the harmonies of nature are turned to discords."[44] He assumed that indigenous peoples and the inhabitants of the countryside lacked the education and foresight necessary to be appropriate stewards of the natural world. For Marsh the state, armed with technical and administrative expertise, was the sole body that could guarantee public environmental goods in a society dominated by short-term private gain. One finds echoes of Marsh's perspective in the records of imperial foresters and land-use managers in India, China, Africa, Indonesia, and Latin America. These officials often complained that customary swidden agriculture, wood gathering, and hunting endangered the "environment," or, more precisely, colonialist schemes to establish forest reserves for timber export.[45] In these societies it was indigenous resource users, not bourgeois wilderness advocates or sportsmen, who provided the most potent opposition to the technocratic ambitions of the colonial or postcolonial state.[46] Conservationists and preservationists of the early twentieth century appealed to Romantic nature as a font of authenticity and an antidote to an overly civilized urban existence, but simultaneously laid the foundation for a thoroughly modern and managerial approach to the natural environment and its human inhabitants.

A Postnational Era, 1945–present

There is little doubt that World War II, particularly the detonation of the first atomic bombs in the 1940s and the threat of radioactive fallout from testing thereafter, galvanized public awareness of the interlocking webs of life that could magnify isolated cases of radioactive poisoning higher up the food chain and across vast distances.[47] In addition, the war's destruction gave new life to calls for international cooperation that included environmental affairs; after all, the issues of greatest concern to environmentalists in the 1950s and 1960s—the spread of carcinogenic pesticides, toxic water pollution, and the destruction of wildlife habitat—knew no national or regional boundaries. Whether this "Age of Ecology" has contributed to a "postnational" phase of environmental history, however, is doubtful. To be sure, there are numerous cases of transnational cooperation yielding environmental improvements in the postwar era. In the case of the Rhine River, for example, an international commission composed of representatives from the Netherlands, Germany, France, Luxembourg, and Switzerland helped to transform the "sewer of Europe" from the 1960s into "the cleanest river in Europe" by the 1990s. The International Commission for the Protection of the Rhine Against Pollution devised clean-up targets for such industrial scourges as chloride salts, PCBs, and thermal pollution that had decimated the Rhine's indigenous salmon and invertebrate populations. Such negotiation proceeded slowly until the Sandoz Chemical Factory spill of 1986. In this so-called Bhopal on the Rhine, water used to put out the flames of the fire washed thirty tons of pesticides, fungicides, dyes, and other toxic chemicals into the Rhine; half a million fish died instantly.

The incident prompted widespread demonstrations and government pressure, spurring the Rhine Commission to negotiate a new set of pollution control standards, the Rhine Action Plan. The plan committed the five nations to ensuring the Rhine's continued use as a source of drinking water, to cleaning up sediment in the river that had become polluted by heavy metals and toxic chemicals, and even to restoring the main stem of the river as a habitat for salmon and other migratory fish.[48] In fact the 1989 Rhine Commission study, *Ecological Master Plan for the Rhine*, or *Salmon 2000*, laid the groundwork for some of the most ambitious ecological restoration work in the world, including the regeneration of alluvial floodplains, the creation of spawning beds, and the installation of fish ladders around weirs and dams.[49] So far, however, *Salmon 2000* has failed to attain anything remotely close to its projected goal of six thousand to twelve thousand adult individuals running annually. Far from it: as of 1999, only 180 had been spotted. The habitat restoration projects have been too small in scope to bring about a significant restoration of riverine plant and animal life, let alone a full rebound of salmon, stork, or water bugs. Nevertheless, they represent the first important steps in reestablishing the river's lost floodplain and alluvial corridor.[50]

European efforts at riparian restoration, moreover, cannot compensate at a global scale for accelerating development, including new hydroelectric dams in such emerging economies as India and China. China's Three Gorges hydroelectric dam complex on the Yangtze River, which was masterminded by former premier Li Peng and finished in May 2006, provides a stunning antithesis to the deterioration and even breaching of dams in other parts of the world. One historian has aptly called it a "vestige of Soviet style central planning by specialists who disdain the opinions of affected citizens," as it displaced hundreds of thousands of residents and submerged hundreds of towns, villages, ancient temples, tombs, and fortresses. The dam also threatened to channel toxic waste into the growing reservoir.[51] The Three Gorges project is the latest example of developing nations' preference to embrace, rather than leapfrog over, the environmentally questionable infrastructural projects first developed in Europe and North America.

Even where developing nations have chosen a more sustainable development path, they have often ignored local subsistence needs and traditional cultural landscapes in favor of attention-grabbing designs that deflect attention from more pressing ecological concerns. Such is the case at Dongtan, a giant, marshy island on the Yangtze River near Shanghai, which garnered worldwide attention in the early 2000s as the world's first planned "eco-city." Designed by "eco-engineers" and green-minded architects and urban planners at the international consulting firm McKinsey and Company and the British architectural firm Arup, Dongtan was heralded as a zero-carbon, car-free, water-recycling model for the world. It was supposed to house twenty-five thousand residents by 2010, just in time for the Shanghai World Expo, whose theme "Better City, Better Life" encapsulates the desire to forego traditional industrialization in favor of a "great leap forward" into brighter ecological modernism.[52]

As of mid-2009, however, only a dozen wind turbines and an organic farm have been completed. Dongtan has fallen victim to corrupt officials, unclear funding priorities, and big-name Western firms who failed to understand the needs of local inhabitants or even notify them about the impending development.[53] In 2007, Ethical Corporation, a Web site on responsible business practices, called Dongtan a "Potemkin village" and "a masterpiece of greenwashing" that was never intended to become a reality.[54] The environmental journalist Christina Larson has described Dongtan as a "model of bold visions that mostly stayed on the drawing boards—or collapsed from shoddy implementation. More often than not, these vaunted eco-cities have been designed by big-name foreign architectural and engineering firms who plunged into the projects with little understanding of Chinese politics, culture, and economics—and with little feel for the needs of local residents whom the utopian communities were designed to serve."[55] Green architects and ecologically informed planners, in this sense, are just as likely as their cameralist and high modernist forbearers to ignore Scott's *mētis*, the informal, on-the-ground social practices that sustain environment and society at local scales.

On Rats, Sharks, and the Nation-State

The dubious environmental history of the modern state thus reveals critical turning points in which European, North American, and later developing nations shifted their ecological systems, economic modes of production, and cultural values in an effort to render both nature and society legible, rationalized, and open to manipulation. Environmental critics of the state, including peasants fighting for customary rights, Romantics bemoaning the materialist debasement of divine nature, or conservationists battling the impoverishment of biological diversity, pointed out the devastating ecological and social consequences of these processes, but until the late 1970s were routinely ignored due to the state's drive to conquer nature and to steer its subjects or citizens away from subsistence modes of production and toward more economically "productive" activities. In recent years, moreover, efforts to develop an ecologically sustainable economy have produced a new cadre of ecological planners who often ignore local needs and citizen input in favor of headline-grabbing but environmentally dubious designs.

Such precarious technological manipulation has left human populations vulnerable to further catastrophes. As J. R. McNeill has written in *Something New under the Sun*: "We, as a species, are unwittingly choosing a particular evolutionary gambit. In the very long view of biological evolution, the best survival strategy is to be adaptable, to pursue diverse sources of subsistence—and to maximize resilience. This is because in the long run there will be surprises, shocks, and catastrophes that kill off some species, no matter how well adapted they may be to one specific set of circumstances."[56] *Homo sapiens,* much like the rat, used to be a master of resilience and adaptability. However, the modern nation-state, as a complex web of government institutions, social actors, customs of rule, and fixed territoriality, has pushed us toward what McNeill has called a "shark strategy," one of "supreme adaptation to existing circumstances." Such fine-tuned adaptation is rewarded by continuous success and can work for millennia so long as governing conditions remain the same. But once conditions change—of greatest concern is global warming and its accompanying hurricanes, floods, and diminishing water supplies—it will inevitably strain the existing social order, including the ideal of a technocratically managed nation-state.

6

Urban Environmental History

JOEL A. TARR

        Urban environmental history is a new subfield that evolved out of the linking of urban history and environmental history.[1] In its simplest form, it concerns the character of the urban environment and the environmental phenomena occurring in cities. Cities historically have grown and expanded over time, developing expansive metropolitan areas and intensive relationships with their hinterlands. Some cities have also experienced substantial decline, especially during the past few decades. Both growth and decline have had sweeping environmental effects. The city's ecological footprint extends not only into its contiguous hinterland but can also leap hundreds and even thousands of miles through trade and communications.[2]

The extent of these impacts depends on a combination of natural and anthropogenic factors. Among the critical natural ones are climate, wind patterns, topography, soil characteristics, and hydrology. Anthropogenic factors include population growth, territorial expansion, industrial development and deindustrialization, changing energy and resource use, and construction of a built environment. All of these factors are affected by the actions of individuals, groups, corporations, and governments pursuing various ends, including economic growth, land development, and the acquisition of power. The actual events that impacted the environment, either negative or positive, include both purposeful actions and unintended effects. Factors such as threats to the public health, various nuisances (including water, air, and land pollution), depletion of natural resources, as well as major natural events drove public attention to urban environmental considerations. The identification of significant turning points depends on such elements as changes in technology and land use, value preferences, theories of disease etiology, the perception of risk held by urban elites and the general population, the costs of change or remediation, and public policies that caused or mitigated environmental problems.[3]

The turning points and events identified here relate primarily to the environmental history of U.S. cities but similar factors have driven change in Europe, especially in the more industrialized nations, albeit at a somewhat different pace and occasionally with different outcomes.[4] American urban history itself can be divided into four overlapping periods. These periods are those of the compact walking city (c. 1790–1870); the industrializing networked city (c. 1870–1920); the development of metropolitan areas (c. 1920–1970); and the era of spreading and splintering urbanization (c. 1970–2000). These are not fixed or absolute time periods; rather, they conform largely to the changing spatial and economic character of cities shaped primarily by economic development, public and private infrastructure construction, and transportation and communications technologies. In each period it is possible to identify distinctive environmental patterns. Cities often faced similar issues revolving around the need to meet their requirements for water, clean air, land, materials, and energy in the context of population, spatial, and economic growth. However, the extent to which problems were solved or mitigated, or even received consideration from a policy perspective, varied sharply from period to period.

The Environment of the Compact Walking City, 1790–1870

America's first urban network developed along the Atlantic seaboard. Its second urban network extended along the inland waterways of lakes and rivers, which was eventually linked with the coastal cities by canals and railroads. Cities in both networks were primarily commercially oriented, largely compact in size, and densely populated. Urban population grew from 202,000 in 1790 to more than 15 million in 1870. In 1790 no city had a population over 50,000 but in 1870, the population in twenty-five cities exceeded that figure, with three more than 300,000 and one city close to a million. Still, except for a few major cities (New York, Chicago, Philadelphia, and St. Louis), those in 1870 were still largely compact rather than spread out.[5] They had limited public transportation systems (hacks, omnibuses, and horse cars) and provided few municipal services. The initial confined size of most cities resulted in relatively small environmental impacts, but population increases and spatial growth, as well as the extension of their ecological footprints, altered this situation.[6]

Americans founded cities in locations where nature offered various attractions, such as on coastlines, where the land's natural contours created harbors; on rivers and lakes that could be used for transportation, water supplies, and waste disposal; and in fertile river valleys with extensive food, timber, and animal resources. Nature provided benefits but also caused many of the annoyances of daily urban life, such as bad weather and pests and more devastating natural disasters (floods, hurricanes, and earthquakes). Urbanites strove to protect themselves against these natural furies, but often their own actions—building on flood plains and steep slopes, under the shadow of volcanoes, or in earthquake-prone zones—increased their danger from natural hazards.[7]

To make their sites more amenable to development and to extend them territorially, urban builders and entrepreneurs often reshaped natural landscapes; they leveled hills, filled valleys and wetlands, and created large areas of "made" land. On this new land they constructed a built environment of streets, houses, factories, commercial structures, office buildings, and churches. In the process they altered urban biological ecosystems, killing off animal populations, eliminating native species of flora and fauna, and introducing new and foreign species. Thus urbanites constructed a built environment that replaced the natural environment, creating a local microclimate, with different temperature gradients and rainfall and wind patterns than those of the surrounding countryside.[8]

The most significant environmental turning point in this period following the creation of a built environment came with the construction of water systems to bring cleaner and more copious amounts of water into the city. Without adequate supplies of clean water, urban survival and growth is not possible. Ample water supplies were required for domestic and industrial purposes, to cleanse city streets, to meet commercial and industrial needs, and to fight fires. Until well into the second half of the nineteenth century, most urbanites obtained their water supplies from wells and pumps, drawing on groundwater and from local waterways. Beginning with Philadelphia in 1801, however, larger cities began to develop centralized waterworks. These systems usually used steam-driven pumps to draw water from various sources for storage in reservoirs and gravity distribution from water towers. Many of the initial waterworks were privately owned, but by 1860 the sixteen largest cities had municipally owned systems, as private companies failed to provide the water resources for various urban needs relating to firefighting, urban growth, and sanitation. Such riverine cities as Philadelphia, Cincinnati, and Pittsburgh built waterworks that drew from neighboring rivers, while others, such as Boston and New York, built systems that took water from upcountry watersheds via aqueducts. Both methods of obtaining water extended the ecological footprints of cities, with the establishment of upcountry reservoirs having the most extensive and far-reaching effects on the city's hinterland.[9]

Piped-in water was usually first provided to commercial and elite living areas that could pay for services, while working-class neighborhoods were supplied through street-corner spigots or wells. The limited water supply in poor neighborhoods provides an early example of an environmental justice issue. The ability of cities to provide improved water supplies for various purposes, even though not equally distributed to all of the city's population, was a major step in improving the habitability of cities.[10] Closely linked to the provision of improved water supplies was used water disposal, especially that contaminated with human wastes. Here technology and system building lagged behind that of water supply. Urbanites in this period usually deposited their wastes in privies and cesspools. These were often poorly maintained, and wastes frequently leaked into groundwater, polluting water sources and endangering the health of urbanites. In many cities surface conduits intended mainly for storm water lined major streets. Some underground sewers existed, usually in commercial areas, and again mainly for stormwater. These usually lacked self-cleansing characteristics and became "sewers of deposit" rather than of removal. Many cities forbade the placing of human wastes in them. Solid wastes and garbage were placed on open dumps, burned, or deposited in waterways. Garbage collection was inefficient, and decaying wastes, dead animals, and horse manure often littered the streets.[11]

Although environmental conditions in the compact walking city were poor by today's standards, the populations and governments often followed "sustainable" strategies (using today's terminology) to deal with wastes. Farmers and scavengers, for instance, collected horse manure from streets and livery stables and human wastes from privies and used these wastes to fertilize garden crops on nearby farms that were often sold back in the city. Scavenging pigs helped keep streets clean by eating refuse and served as meat for working-class people. Horses commonly died on city streets, but scavengers usually quickly removed them to rendering factories because they had value as a recyclable commodity. City councils passed nuisance statutes to force such polluting industries as slaughterhouses, rendering plants, and coke ovens to move to the urban periphery. As railroads entered the city centers in the 1830s and 1840s, because of the risk of explosions and the disruption of traditional street life, city councils often banned steam locomotives from downtown streets and required that horses be used to pull them to the train stations. These "sustainable" measures largely disappeared near the end of this period, however, as cities grew in population, extended their built environments, industrialized, and expanded their bureaucracies and services.[12]

Public health concerns drove various attempts to clean the streets. Epidemics of cholera, typhoid, typhus, and yellow fever frequently ravaged American cities during this period. Various hypotheses existed about epidemic causation, but those that came to dominate medical thought can be summarized as contagionist and anticontagionist. Physicians who believed in contagionism held that epidemics were spread by human contact, while anticontagionists held that decaying organic matter that gave off miasmas (gases and smells) generated them. Contagionists argued that the best way to protect a city from contamination from epidemic disease was to impose quarantine while anticontagionists maintained that cleaning the streets and homes of decaying organic matters (such as manure and dead animals) would control epidemics. City health departments, where they existed, often endeavored to clean and flush streets, empty privies, and improve sanitary functions. The increased activities of municipal governments in the realm of public health as well as in supplying clean water were important turning points in terms of state intervention in society for health-related purposes, often connected to the environment.[13]

Air quality depended on available fuels as well as on local topographical conditions. Wood and coal were the primary fuels and contaminated the air with their smoke. Manufacturing processes—charcoal making and coke making, rendering and slaughtering—gave off heavy smoke and noxious odors, and some towns and cities banned them from their locales.[14] All cities, in their quest for fuel and construction materials, exploited and reduced the forested resources on their sites and in their hinterlands, again extending the city's ecological footprint in a manner destructive to the environment. From an environmental perspective, the most significant turning point involving air quality was the increasing use of mineral fuels, particularly anthracite and coal, to replace wood and waterpower for industrial, transportation, and domestic purposes. The mining of mineral fuels also had serious environmental effects, disrupting the landscape, polluting groundwater and local watercourses with acidic mine drainage, and causing large fish kills. While anthracite is essentially a clean fuel, bituminous (or soft) coal is exceedingly dirty, and emits sulfur, carbon dioxide, and particles when burned. As American cities became more dependent on bituminous coal for fuel, their skies became increasingly overcast, depriving urban populations of the light and health-related rays provided by the sun.[15]

During the period of the walking city, except for several major cities (New York, Boston, Chicago, and Philadelphia), distances remained relatively small and densities high. To facilitate expansion, some cities took steps to expand their landed areas by filling in adjacent wetlands and waterways. The most extensive case of such site enlargement involves Boston, founded in 1630. The Boston site was originally composed of several small peninsulas and islands, surrounded by shallow tidal flats. As the city grew in the early nineteenth century, the municipality provided the land for shipping, railroad facilities, and residences by filling in the tidal flats surrounding the peninsula. The city-building process also provided soil from the leveling of hills and from cellar dirt excavated in the construction of new buildings. During these years Boston created more than ten thousand acres of land. Other cities, such as New York, New Orleans, and Chicago, as well as Boston itself, obtained the land they needed to grow through a series of major alterations of their topography and by creating new land. In the process these cities caused extensive environmental damage, destroying thousands of acres of marshes (wetlands) and altering the flow of rivers and estuaries.[16]

A second major alteration in the use of city land during this period involved park creation. Up until the 1830s and 1840s, American cities had extremely little open space. In the 1830s, however, rural cemeteries appeared in such cities as Boston and Brooklyn, providing resting places for the dead outside of crowded inner-city churchyards and places for rest, repose, and enjoyment of nature by urban populations. In addition to the rural cemeteries, some cities began to create parks, the most significant of which was New York's Central Park. This park of 843 acres (341 hectares), sculpted by Frederick Law Olmsted and Calvert Vaux, became a model for other urban green spaces. During the next few decades Olmsted and his firm constructed a number of urban parks throughout American cities, opening up the possibility for urbanites to interact with nature. Even though these parks initially primarily served the elite, they eventually became places used by the city's broader population, providing the possibility of interaction with nature for many urban dwellers.[17]

In the period of the walking city, cities had to cope with a variety of environmental issues. The primary ones related to water supply, wastewater and solid wastes disposal, noxious nuisances, public health, open space provision, and site modification. Most municipalities were initially poorly prepared to deal with the major environmental and infrastructure issues that confronted them in a stage of rapid urban development and technological change, although considerable learning took place over the decades. In this context of rapid development, improvements in water supply, site alterations, and the provision of open space can be identified as the most significant environmental turning points of the period, involving both the gaining of environmental improvements and environmental losses.

The Environment of the Networked City, 1870–1920

The period between 1870 and 1920 was one of rapid urbanization in the United States. The number of persons living in cities increased from about 15 million in 1870 to 25 million in 1900 and to 54 million in 1920. By that year there were forty-three cities with populations between 100, 000 and 250,000, thirteen with 250,000 to 500,000, nine with 500,000 to 1,000,000, and three with more than a million. Not only did the urban population and the number of cities increase dramatically, but the urban network, which had been located almost entirely east of the Mississippi in 1860, now extended from coast to coast. In this period forces of concentration and deconcentration were at work to change the structure of the compact walking city. Cities expanded spatially, as their core areas began to assume increased central business district (CBD) functions. Neighborhoods surrounding the business districts attracted dense concentrations of workers, while such cities as Chicago, Pittsburgh, and Toronto experienced the growth of working-class suburbs around outlying industries. Members of the middle and upper classes began moving out of the city's core and into residential suburbs.[18]

Water supply and sewerage, both key elements in the city's metabolism, continued to have major effects on the urban environment. The number of water works in the nation increased from 1,878 in 1890 to almost 10,000 in 1924, while the U.S. population served by sewers increased from about 16 million to more than 47.5 million during approximately the same period. The provision of running water to households resulted in the expansion of the use of domestic water-using appliances. In turn, the output from these appliances increased the flows of gray and black wastewater that needed disposal. The water closet, which discharged water polluted with fecal matter, ranks high among the major environmentally altering technologies of the era, an unpredicted occurrence. Older waste receptacles like privies and cesspools were inadequate to handle the larger wastewater flows. Eventually, cities constructed sewerage networks to accommodate the increased volumes of wastewater, but usually well after the introduction of running water. These sewers transformed the hydrology of urban areas, since sewer pipes followed natural drainage lines and the systems often incorporated streams into pipes and culverts. Thus urban streams ceased to be an element in a site's ecology and became a part of the city's infrastructure network, an important turning point for nature in the city.[19]

Although sewers produced important health benefits for urbanites by eliminating overflowing privies, sewage discharges frequently caused negative environmental and public health effects. During these years most cities discharged their untreated sewage into neighboring water bodies (a small number operated sewage farms). In 1909, for instance, the raw sewage of 88 percent of the urban population was discharged into neighboring waterways. The sewage frequently destroyed the fish life and stream ecology of the receiving water body by absorbing its oxygen. Most critically, the untreated sewage produced serious health costs for downstream cities that drew their water supplies from the receiving rivers.

Mortality and morbidity rates for such infectious diseases as typhoid soared in cities that discharged their untreated sewage into water bodies from which they drew their drinking-water supplies or into which upstream cities discharged their sewage. The provision of running water to cities had been a major benefit, but now its negative effects loomed larger as city after city experienced typhoid epidemics as well as health problems from other waterborne diseases. In 1900, for instance, Pittsburgh had a typhoid fever mortality per 100,000 of 144, Richmond 104, and Washington, D.C., 80. These were all cities that drew their water supplies from neighboring rivers. In contrast, cities that drew their water supplies from protected upstream watersheds, such as Boston and New York, had much lower rates.[20]

A justification for the discharge of raw sewage into streams was provided in 1912 by the Superintendent of the Pittsburgh Bureau of Construction, N. S. Sprague, when he commented: "Rivers are the natural and logical drains and are formed for the purpose of carrying the [city] wastes to the sea."[21] This reflected his belief that nature existed for the benefit of humankind, a position widely shared by other engineers at the time. Although the number of American cities that treated their sewage actually advanced over the following decades, primarily to avoid nuisance, the great majority continued to discharge raw sewage into neighboring water bodies because of its cheapness and convenience in terms of disposal and because sanitary engineers argued that the process of dilution would dispel the wastes. Thus these waters were conceptualized as "a kind of 'organic machine' for the processing of human waste," the social utility of which rested on changing concepts of health risks and other water resource uses.[22]

Aside from shifting a city's water supply to a protected upstream source, the primary solutions to the threat of waterborne disease were to filter and chlorinate the water. These technical and chemical approaches developed in the late nineteenth and early twentieth century and resulted in major health advances for cities whose water supplies had been contaminated by sewage. Public health officials also advocated sewage treatment to provide a second safety net for drinking water supplies and to avoid nuisances, but unless faced by obvious problems, cities preferred to rely on dilution for disposal rather than incur the cost of a sewage treatment plant.

Many cities followed the practice of filtering their water but not treating their sewage. Raw sewage in many urbanized areas overwhelmed the oxidation capacity of the rivers, creating offensive sights and smells. Water treatment had sharply reduced typhoid deaths, but diarrhea and enteritis death rates remained elevated. Fish were absent from long dead stretches of the rivers and chemical pollution fouled the taste of many drinking water supplies as urban waste disposal in waterways created an ecological footprint that threatened the health of downstream cities.[23] At the core of the city the development of the CBD or downtown had serious environmental consequences. The CBD housed the control panels of the growing urban business, transportation, and communications networks and provided both an aboveground and belowground example of the substitution of the built environment for the natural environment. Business district streets in the larger cities were often lined with multistoried structures, and even skyscrapers, after the 1880s and 1890s. Streets were paved with impermeable materials, covered with streetcar tracks, and arched-over by electrical, telephone, and telegraph wires. Water and sewer pipes and in a few cases subways and freight tunnels extended deep under the ground, penetrating geological strata that had not been disturbed for eons. These dense urban areas developed their own climatic conditions, creating urban heat islands and wind tunnel effects.[24]

Suburbanization reshaped the patterns of the compact and congested walking city and expanded the built environment over larger areas. Members of the elite and of the middle class, many of whom had earlier lived in the core, moved to the city's outlying areas and suburbs, seeking rural amenities, cheaper land, and escape from the city's noise and congestion. Improved transportation—especially electrically powered streetcars—enabled the rise of streetcar suburbs and greatly expanded the areas within commuting distance of the downtown core. From 1890 to 1907 track mileage increased from 5,783 miles (9306.8 kilometers) to 34,404 miles (55, 367.9 kilometers) and rides per inhabitant to 250, as the use of public transit far outdistanced urban population increase. From an environmental perspective, the expansion of the city into its hinterlands resulted in the conversion of land that had been farms, forests, and open space into the built environment of residential suburbs.[25]

The decentralization of industry extended the ecological footprint of urban industries deep into the city's hinterland. Numerous firms moved to the urban fringe along railroad lines, waterfronts, and rivers, and in the process they altered the land to facilitate production, controlled rivers for commerce through locks and dams, and hardened and shaped riverbanks to meet their transportation needs. Increased industrial production greatly accelerated the volume of wastes contaminating urban air, water, and land. Industrial wastes therefore began to rival human sewage as a major cause of the degradation of the urban environment.[26]

The Progressive Movement occupied the latter part of this period, bringing many attempts to improve, reform, and order American society. It produced significant improvements in urban life, including the reduction of corruption in city governments, regulation of housing and sanitary conditions, and advances in public health. Reform groups, many led by women, paved the way in driving government to improve urban environmental conditions. In addition to the important changes in water-supply quality, attempts were made to clean the streets, control smoke, and reduce noise. Many of these reforms, however (such as smoke and noise reduction), had only narrow effects in the face of limited governmental action and population and industrial growth.[27]

During the rise of the networked city, there were several critical urban environmental turning points. As cities grew in size, municipalities provided increased amounts of infrastructure and services. But although these infrastructure systems often brought public health improvements and reduced nuisances, they transformed natural aspects of the city site and generated significant environmental and public health problems. Especially severe were the health costs imposed on downstream cities by the sewage disposal practices of upstream cities. In this period the rate at which cities extended their ecological footprints into their nonurban hinterlands accelerated, although it remained limited compared to the following city stages. Industrial growth also accelerated, increasing the burden of polluted air and water emissions on the urban environment.

Cities and Suburbs in a Metropolitan Context, 1920–1960

From 1920 to 1960 the U.S. metropolitan population increased from approximately 36 million (35.7 percent of the total population) to more than 113 million (63.3 percent). In 1920, 9.2 percent of the metropolitan population was suburban, but in 1960 the percentage had risen to 30.9 percent. In contrast to their suburbs, some large eastern and midwestern central cities experienced relatively little growth after 1930. However, many of those in the West and the so-called

Sun Belt states grew dramatically after World War II, increasing their burdens on the environment. In these years, aside from interruptions during the Depression and World War II, cities spent increased efforts addressing such environmental issues as water quality and solid waste disposal compared with previous decades, although many environmental problems persisted. Most cities disinfected their water supplies to protect the public health against infectious disease and extended their sewer systems citywide, but they only treated their sewage when it created severe nuisances. These were actions primarily taken by municipalities, but during the New Deal federal programs provided for improvements in urban water supply and sewerage systems as well as many city parks. Still, results were limited, and as late as 1950 more than half of urban sewage was disposed of in waterways without treatment.[28]

Some advances were made in regard to solid waste disposal. Many cities motorized their collection fleets and some experimented with disposal by incineration before World War II. Most, however, continued to rely on open dumps and ocean dumping or pig farms, for disposal. After the war both cities and suburbs embraced the new technique of the sanitary landfill. Sanitary engineers originally believed the sanitary landfill to be the solution for solid waste disposal, but by the end of this period its limitations in regard to pollution of the groundwater, subsidence, and other problems were becoming obvious. During the same postwar decades, the development of such new products as plastics, major increases in consumption and waste generation, and the appearance of a wide-range of innovative chemical products (including pesticides and herbicides with poorly understood toxic characteristics) resulted in landfills with greater toxic hazards than earlier ones devoted to organic wastes and trash.[29]

A huge growth in automobile usage marked these decades, providing a major turning point in the acceleration of suburbanization and its land-consumption patterns. Automobile ownership exploded after 1920, and by 1930 there was an automobile for every 5.3 persons in the United States. By 1960, 77 percent of households owned an automobile, with 15 percent owning two or more. The automobile extended the suburban boundary into greenfield areas beyond the earlier suburbs and filled in the undeveloped land left between streetcar tracks. Commuter traffic soared, as did downtown congestion, with many drivers commuting between bedroom communities and downtown jobs.[30] State and local governments paved millions of miles of roads and constructed thousands of miles of new highways, extending the city's footprint. The automobile and its supporting infrastructure had extensive environmental impacts. One was the replacement of the horse by the internal combustion engine, thereby freeing city streets of the nuisances and health hazards presented by horse manure, a seemingly positive effect for the quality of the urban environment.[31] Many others, however, were negative, including increased air emissions of pollutants and state and local government paving of millions of miles of roads and construction of thousands of miles of new highways. The interstate highway system, authorized by Congress in 1956, was important in facilitating and accelerating the move to the suburbs. The new suburbs themselves consumed thousands of acres of greenfields and generated additional wastewater and solid waste streams requiring disposal.[32]

A major positive step involved improvements in urban air quality, an odd development considering the increases in automobile use. Air quality concerns throughout the first half of the century had focused on smoke from coal-burning industries, residences, utilities, and railroads. The absence of any national or state air-quality regulations meant that this issue would be dealt with at the city and county levels. The driving forces behind the smoke control movement were a desire to clean city air and increase sunlight, to diminish the negative effects of smoke on health, to reduce nuisances such as the soiling of clothing and building facades, and to preserve vegetation. Civic leaders additionally saw smoke control as a necessary strategy to prevent the population from fleeing the smoky city for the suburbs. Such industrial municipalities as Cincinnati, Pittsburgh, and St. Louis therefore enacted stringent smoke control ordinances, accelerating a movement toward the use of cleaner fuels like natural gas and petroleum. Serious air pollution episodes, such as the Donora smog disaster of 1948, also heightened attention to the hazards of industrial waste emissions but did not produce any significant regulation.[33]

One consequence of smoke control was the transfer of some of the negative impacts of coal emissions to localities distant from the city. Although many fuel consumers substituted oil or natural gas for coal, there were other large users (such as electrical utilities) that increased their coal consumption. Many of the municipal smoke control regulations focused on perceptible ground-level pollution, and in the 1950s and 1960s air pollution control experts advocated tall stacks as a means to shift coal emissions into higher atmospheric levels, where they would supposedly dissipate. By the 1970s, however, scientists understood that coal emissions underwent chemical transformations in the upper atmosphere and were deposited as acid rain in areas distant from the cities. Thus a technology devised to help rid cities of their local pollution—tall stacks—had extended the city's ecological footprint in a manner similar to the impact of the discharge of sewage into streams on the water supplies of downstream cities.[34]

Metropolitan growth consisting of the rise of large central cities with growing suburbs stimulated new environmental problems as well as exacerbated old ones. Cities still struggled with issues of wastewater and solid waste disposal made more formidable by growing populations. New suburbs often experienced difficulties in dealing with similar issues.[35] And political boundaries did not halt transboundary pollution, be it air or water, so that the environmental burdens created in one jurisdiction could affect other locales. Some New Deal measures provided funds to improve such urban infrastructures as parks, sewers, and water supply systems, but financially strapped cities did little on their own. Congress passed no federal legislation protecting the environment from pollution, and although some states began regulating water pollution in the first third of the twentieth century, river pollution, especially in interstate rivers, remained a serious issue. Increasingly it became clear that urban environmental issues could only be successfully managed on a metropolitan, regional, or even national scale, as the ecological footprint of the built environment extended into the surrounding rural and natural environments.

Central Cities and Edge Cities in the Environmental Era, 1960–2000

The rise of the environmental movement in the 1960s and 1970s had strong implications for cities and must be regarded as a major turning point in the history of the urban environment as well as in environmental history in general. As environmental historian Samuel P. Hays has observed, cities were also a "major source of the environmental movement" because they had the "largest membership in environmental organization and the greatest expression of environmental values."[36] Thus, although cities created substantial environmental problems, they also provided a major catalyst to drive environmental improvement. In the 1960s and 1970s Congress and the states enacted a broad-based set of environmental regulations relating to air, land, and water, all of which affected the quality of the urban environment.[37]

The most significant acts for cities related to air and water quality and to solid and hazardous waste disposal. Air-quality concerns were especially strong in automobile-intensive locations like Los Angeles, and in 1947 Los Angeles County created the first countywide air pollution control district. It was not until 1952, however, that Dr. Arie Haagen-Smit discovered that automobile emissions of nitrogen oxides and hydrocarbons in the presence of ultraviolet radiation from the sun formed photochemical smog.[38] In the ensuing decades, various states and counties, following the lead of California and Los Angeles, passed air pollution control legislation of varying severity. In 1963 Congress enacted the first Clean Air Act and other acts followed in 1967, 1970, and 1977, setting clean air standards. In addition, the Environmental Protection Agency (formed in 1972) moved to phase out the use of leaded gasoline beginning in 1973. These acts included specific technology-forcing provisions, reflecting the conviction that only incentives would push some firms to install innovative and environmentally beneficial technologies.[39]

Regarding water and waste disposal, the Clean Water Act of 1972 set national goals for water pollution control, and the Safe Drinking Water Act of 1974 directed the EPA to set maximum contaminant levels. EPA enforcement of these acts resulted in a significant cleanup of urban waters, largely supported by federal funds to construct sewers and sewage treatment facilities and to upgrade water supply systems, although they fell far short of the goal of the Clean Water Act to achieve "zero discharge" by 1985.[40] Solid waste problems were also reduced in the 1970s, when the EPA moved to end open dumping, sought source reduction, and advocated improvements to operating criteria for sanitary landfills. Congress also passed the Resource Conservation and Recovery Act of 1976 to provide incentives for recycling and recovery of waste materials.[41] Pollution of the land and groundwater, however, had escaped close scrutiny, until public outrage over the discovery of several sites badly contaminated with toxic substances in the late 1970s—the most famous of which was Love Canal in 1978—resulted in Congress passing the so-called Superfund legislation in 1980 (CERCLA) and 1986 (SARA). This legislation was intended to deal with former industrial sites contaminated with hazardous substances and constitutes a major turning point in the history of the urban environment because of its unprecedented focus on land contamination. Eventually, an EPA inventory listed thousands of "brownfields" located throughout the nation, largely in metropolitan and inner-city areas.[42]

As a result of the congressional legislation throughout the 1970s, a number of urban environmental problems—such as sewage pollution of streams, degraded urban water systems, poor air quality, and open dumping on the land or in waterways—were mitigated. Although these acts provided large improvements in the urban environment, they were vulnerable to political and bureaucratic manipulation. Political questions in the 1980s about the benefits and costs of environmental legislation, as well as opposition from industrial interests and developers, led to a reduction of federal funding for such programs in the 1980s, resulting in slower movement toward higher levels of urban environmental quality in the following years.[43]

The most significant technology affecting urban environmental conditions continued to be the automobile. The growth in automobile use during these decades was very large, and even though emissions per vehicle were reduced, air-quality standards were frequently exceeded. Between 1945 and 2000 automobile registrations increased from 25,796,000 to 132,247,286. In 1950, 52 percent of households owned an automobile and 7 percent owned two or more; by 2001 the number of vehicles per household was 1.89, or almost an average of two vehicles per household. Average trip length and the number of vehicle trips taken per person and per household surged, with a decline in other modes of travel besides single-occupancy driving. Even though highway mileage expanded, automobile use and hence congestion outgrew it.[44] A prime factor in facilitating the increase in automobile use was the availability of new limited-access highway systems. Although some states had such highways in the late 1930s and in the initial postwar period, the Interstate Highway Act of 1956 greatly facilitated automobile commuting and transport. Under the act the federal government agreed to distribute its 90 percent payment of interstate expenses according to local needs. Approximately half of the money spent on highway construction under the act went to urban areas.[45]

In the decades after World War II, suburbanization increased at a rapid rate, supported by such government policies as the Veterans Bill and the tax deductibility of mortgage interest as well as the improved access provided by the new highway systems. The mechanization of housing construction, most strikingly displayed in Levittowns outside of Philadelphia and in Long Island, resulted in massive suburban developments that builders soon replicated in most metropolitan areas. By 1980 suburban population had surpassed central city population.[46] Another major change in metropolitan land-use patterns developed primarily in the 1980s and 1990s. This was the expansion of suburban growth on the urban fringe and the creation of the new "outer" or "Edge City." The outer city replicated many of the elements that had made downtowns so attractive a century before, including shopping, entertainment, banking, and commercial facilities in one location. Unlike the city downtown, however, these shopping and business nodes were primarily automobile accessible, provided ample parking facilities, and were linked to major highways. Usually, they were poorly served by public transit. Such development was also aided by the growing proliferation of information technology, permitting distances between economic and information functions to increase.[47]

In addition to absorbing greenfield areas, massive suburban residential building and the construction of highways and shopping malls led to species destruction, increased erosion, altered hydrologic patterns, and pollution of rivers and lakes. In addition, many suburban developments were inadequately served by wastewater disposal systems and depended on septic tanks for domestic wastes, often causing groundwater pollution. These developments, with their emphasis on consumption, both consumed nonrenewable resources and generated large amounts of waste.[48] From an environmental point of view, further suburbanization and the growth of the outer city resulted in a major increase in land consumption often called sprawl. While the average density of urbanized areas was 2,929 persons per square mile (about four persons per acre) in 2000, the population density on land developed since 1960 is 1,469 persons per square mile (a little over two persons per acre). Even in metropolitan areas where population has stayed static or even declined, land consumption has greatly increased while densities decreased. In contrast, many of the densest metropolitan areas in the United States are located in the West, in such cities as Los Angeles, popularly thought to be marked by low-density sprawl.[49]

Also qualifying as a major urban environmental change are the shifts in regional growth that have accompanied urban decentralization mainly in the so-called Sun Belt. By 1990 forty of the nation's hundred largest cities were in the South, and only thirty in the North. As urban populations shifted to the Sun Belt states, the nation became increasingly dependent on air-conditioning technology and its high-energy demands. In addition, water scarcity in many of these regions and the extensive increases in water use that suburban growth generated put severe pressure on water resources.[50]

While metropolitan areas were becoming increasingly decentralized, many central cities were losing industrial and commercial establishments as well as people, another major turning point. Deindustrialization caused job loss and economic decline as well as leaving urban brownfields often contaminated with toxic substances. Perversely, however, in some regions deindustrialization actually produced improvements in air and water quality by eliminating major emitters of pollutants.[51] From the perspective of land-use practices, the United States is faced by two primary opposite tendencies: the growing urbanization of greenfields on the periphery of metropolitan areas and the persistence of polluted brownfields in the inner city. Brownfields involve not only issues of land use and pollution but also critical questions of environmental justice. Many of the populations residing near brownfields were ethnic and racial minorities as well as the poor more generally. The focus within the environmental movement and within the environmental history profession itself upon issues of environmental justice and environmental racism reflects a shift within these communities to more of a social orientation.[52]

In contrast to these continued environmental impacts is the movement toward urban sustainability that is attempting to halt or modify many of the trends described in this chapter. From the perspective of the urban environment, the sustainable cities movement aims toward a reversal of sprawl patterns, restoration of a sense of community in urban neighborhoods through the use of planned urban design, revitalization of the inner city, a reduction in automobile use, an enhancement of urban public transit systems, and an emphasis on the use of renewable resources and recycling.[53] Whether it will be possible to accomplish these changes has yet to be seen. If the sustainable cities movement succeeds, it will constitute another major turning point in urban environmental history.

THIS CHAPTER HAS spelled out some of the main themes involved in the interaction of American cities and the natural environment and identified major turning points. In doing so, it has used perspectives that have evolved from such concepts as the metabolism of cities, construction of the built environment, changing land uses, and the evolution of the city's ecological footprint. These concepts were explored within a framework of city development encompassing the walking city, the networked city, the metropolitan region, and the outer city. Each period had innovations relating to infrastructure, industrial production, transportation, and fuel type that often resulted in improvements in local conditions but had also generated environmental costs and frequently extended the city's ecological footprint. Public policies to deal with these problems evolved over time, first on the local level, then on the county and state levels, and finally on the federal level, although improvements in urban environmental conditions continue to be uneven. Although some environmental conditions in U.S. cities and suburbs improved as cities moved from stage to stage, new problems resulting from poor planning, a weak knowledge base, deteriorating infrastructure, land-use changes, unpredicted effects of new technologies, and huge increases in consumption have evolved while old ones have frequently persisted. Conflict and change resulting from a clash between the forces of decentralization and centralization make it difficult to predict what the environmental problems or improvements of the next stage of urban development will be.[54]

A Note on Urban Reactions to Natural Disasters

I have spent very little time exploring natural disasters as key turning points in urban history. This is not because I consider them unimportant, but rather because in many cases natural disasters have not resulted in major changes in correcting long-term urban problems nor in guarding against the risk of future disasters of a similar kind. Americans have taken some actions to protect against catastrophes caused by natural events; they have strengthened building codes in response to hurricanes and earthquakes, limited building in some areas, raised the costs of insurance, and built higher levees and dams to protect against floods. In some cases they have capitalized on the opportunity to make improvements on the predisaster city.

Despite these actions, however, and perhaps in some cases because of them, Americans still continue to build on flood plains, in low coastal areas subject to hurricanes and tsunamis, in earthquake-prone locations, and relatively close to live volcanoes.[55] By their very actions they have often increased their vulnerability, making the obvious point that so-called natural disasters have strong social connections. In addition, cities have seldom changed their basic urban patterns in response to catastrophe. As several authors have observed, the power of property rights to stabilize urban forms and to limit their evolution in new directions are immense.

There are exceptions, of course, and one could point to Galveston, Texas, as a city whose future was sharply altered by a major hurricane, but it is difficult to point to another example in the United States. As a recent book on disasters (not all "natural") has observed, in the nineteenth and twentieth centuries, "resilience" has been a "nearly universal fact of urban settlement."[56] The recent case of New Orleans and Katrina may prove the opposite, and innovative developments in housing and neighborhood design are taking place. However, it is too early to know for sure whether New Orleans will deviate significantly from historical patterns.

7

The "1950s Syndrome" and the Transition from a Slow-Going to a Rapid Loss of Global Sustainability

CHRISTIAN PFISTER

❧ During the international geophysical year of 1957–1958, the geophysicist Hans Suess and the oceanographer Roger Revelle, who was the mentor of Al Gore, discovered that the CO_2 content of the atmosphere had risen since it had first been measured in the mid-1890s by Svante Arrhenius. The two scientists framed their finding this way: "Thus, human beings are now carrying out a large scale geophysical experiment of a kind that could not have happened in the past, nor could it be reproduced in the future. Within a few centuries we are returning to the atmosphere and oceans the concentrated organic carbon stored in sedimentary rocks over hundreds of millions of years. This experiment, if adequately documented, may yield a far-reaching insight into the processes determining weather and climate."[1]

Today this statement impresses, considering the turn of events that has since occurred. Revelle and Suess used the term "experiment" in its traditional sense of an opportunity to study (geo-)physical processes. Yet they felt that there might be some risk of climate change from the rise in CO_2, albeit in the distant future.[2] CO_2 concentration, as estimated from ice cores, had only slowly risen from 297 parts per million (ppm) in 1900 to about 316 ppm in 1957, while today's concentration (as of 2010) is 395 ppm[3] (figure 7.1). Based on an extrapolation of the figures from the first half of the twentieth century, this value should not have been reached until 2212![4] This illustrates how close to sustainability the global environment was in the wake of World War II, as compared with today's situation.

By the mid-1960s Revelle had already shared with his students the dramatic

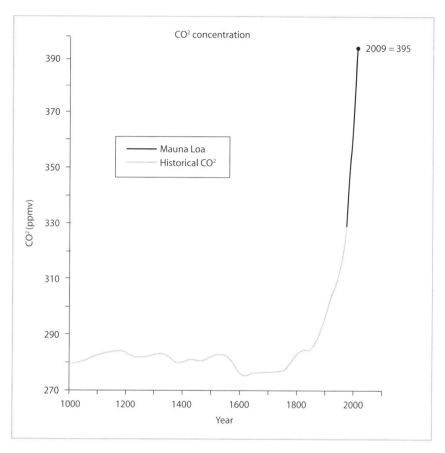

Figure 7.1. The rise of greenhouse gas concentrations in the atmosphere A.D. 1000 to 2010. *Source:* Data up to 1957 is from D. M. Etheridge, L. P. Steele, R. L. Langenfelds, R. J. Francey, J.-M. Barnola, and V. I. Morgan, "Historical CO_2 Records from the Law Dome DE08, DE08-2, and DSS Ice Cores," in *Trends: A Compendium of Data on Global Change* (Oak Ridge, Tenn.: Carbon Dioxide Information Analysis Center, Oak Ridge National Laboratory, U.S. Department of Energy, 1998), available online at http://cdiac.ornl.gov/trends/co2/lawdome.html. Data from 1958 is available on ftp://ftp.cmdl.noaa.gov/ccg/co2/trends/co2_annmean_mlo.txt (accessed on May 13, 2010).

results of the first eight years of CO_2 measurements, as concentrations were rapidly increasing every year. He predicted that a continuation of this trend would force a profound and disruptive change in the entire global climate.[5] In 1981 the renowned meteorologist Hermann Flohn issued the following warning: "In any case, the problem must be taken seriously: it threatens humanity as a whole, and in the course of the coming century it will threaten the generation of our children and grandchildren.... It is not a matter of arguments in the short-term political arena; the fate of our children and grandchildren all over the whole world is at stake."[6]

Most textbooks relate the origins of the greenhouse problem to the onset of industrialization. However, the comparison of the growth rates of CO_2 emissions before and after 1950 suggests that the immediate roots of the greenhouse problem in its present and future urgency are rather to be found in the twenty-five years between the late 1950s and the early 1980s. A trio of renowned scientists—including Will Steffen, the current executive director of the International Geosphere-Biosphere Programme (IGBP); Nobel laureate Paul Crutzen; and the historian John McNeill—looking at the global use of fossil fuels and raw materials over the past two centuries concluded in 2007 that the period since the onset of industrialization needs to be divided in two stages of unequal growth: namely, a first stage of slow growth before 1950 and a second stage of exponential growth since that time. Focusing on the exponential growth rates in the production of a multitude of raw materials since the 1950s, the three authors named this second stage "The Great Acceleration," probably borrowing from Karl Polanyi's concept of "Great Transformation."[7] Emphasizing humankind's focal role as a force in geology and ecology, Crutzen coined the term "Anthropocene" for this period, mirroring the names of other geological epochs such as the Holocene.[8] The trio concluded that economic growth, particularly the "new regime of international institutions after 1945," alongside rapid technological change and population growth, were the main causes of the "Great Acceleration."[9]

This argument about the two-stage character of the industrial age is not a new one. In 1992, inspired by the Meadows diagram of the remarkable take-off shift in global energy use since the 1950s (figure 7.2), I published a short essay in a Swiss environmental journal portraying this surge of energy use as historically unique and relating it to "low prices for fossil energy."[10] In that article I gave the phenomenon the name the "1950s syndrome." A few years later, in collaborative work on a variety of issues—such as the global oil market, road taxation, urban sprawl, air pollution, social change, the environment for children, energy, and policy options—I became convinced that most of these symptoms were in some way affected by one common driving force.[11] My key conclusion was that the decline in the price of fossil fuel since the 1950s, seen in relation to the price of labor and capital, was the most significant cause of the wasteful consumption of raw materials and energy, and the resulting excessive environmental stress.[12] The

The "1950s Syndrome"

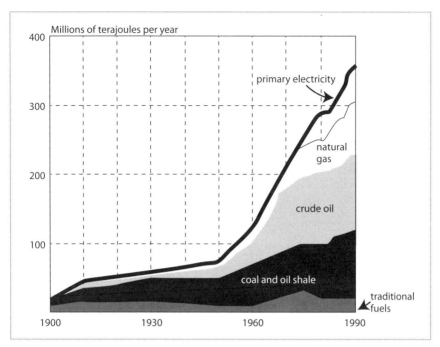

Figure 7.2. World energy use, 1900–1990. *Source:* Donella H. Meadows, Dennis L. Meadows, and Jørgen Randers, *Beyond the Limits: Confronting Global Collapse, Envisioning a Sustainable Future* (Post Mills, Vt.: Chelsea Green Publishing Co., 1992), 362.

"1950s syndrome" appeared in an English summary in 1998.[13] A 2003 essay in a popular treatment of environmental history responded to some objections to the concept, while broadening the argumentation by including the issue of high-input agriculture.[14] For some reason or another, Will Steffen and his coauthors overlooked the "1950s syndrome" altogether.[15]

Reception and Discussion of the Term "1950s Syndrome"

The term emphasizes the 1950s as the significant turning point from a slow-going to a rapid loss of sustainability. "Syndrome" is borrowed from the medical field, where it designates a bundle of indicators, symptoms that collectively describe a disease. It also indicates that these different manifestations of change have not heretofore been studied together.[16] In general the concept of the "1950s syndrome" has been well received within environmental history.[17] According to

Jens Ivo Engels, the term is attractive because it combines physical, social, and cultural aspects.[18] Arne Andersen has adopted the metaphor for his history of consumption.[19] Patrick Kupper has connected it to the fundamental reinterpretation of humankind's relationship with the environment in the early 1970s, for which he coined the term "1970s diagnosis."[20]

Outside the discipline of environmental history, other scholars commenting on the term have been more critical, however. Political historian Axel Schildt has pointed out that the conclusions did not apply to Germany or even to Western Europe.[21] Hansjörg Siegenthaler, one of the founding fathers of Swiss economic history, has argued that cheap energy was not the cause of the boom but rather one of its secondary preconditions.[22] Many reviewers and subsequent readers, among them such renowned German economic historians as Wolfgang König, have echoed Siegenthaler's assessment, claiming that the concept grossly overestimated the significance of cheap energy. However, neither Schildt nor König nor Siegenthaler have dealt with the issue of energy at all.[23]

(Environmental) historians often balk at dealing with prices and including economic arguments in their narratives. This resistive attitude is directed against the neoclassical worldview of "economic man," which is inherently ahistoric. It sees the economy as a closed system within which goods are produced by inputs of capital and labor, and then are exchanged by consumers and firms.[24] The field of ecological economics would be far better suited for an interdisciplinary cooperation with (environmental) historians. According to Peter Söderbaum's brilliant survey, it is open to scholars from ecology and social sciences who share the vision of ecological sustainability, although markets, wages, and prices play a role.[25] There is a distinct and unresolved divide between neoclassical and ecological economists with ecological economists arguing that the neoclassical worldview fails to account for considering the physical basis of industrial production, which is energy.[26] For neoclassical economics, energy is just a cost factor, while for ecological economics it is a "fundamental enabling economic production."[27] At the same time it is known to be the key variable for the analysis of ecosystems. Economist Hans-Christoph Binswanger argued in 1974 that "omitting energy as a production factor is the fatal flaw of both traditional liberal and socialist economics."[28] This fundamental weakness of mainstream economic theory is at the root of the market failures that have caused the current critical global situation.[29]

Does the "1950s syndrome" stand for an epoch-making transition, dividing the period of the "industrial society" from the period of "consumer society"?

Environmental and economic historian Rolf Peter Sieferle and his coauthors have argued for interpreting the very long trend in global economic growth including the post-1950 period as a resumption of an earlier growth phase being interrupted by the exceptional circumstances during the two world wars and the intermediate crises. He also allows for the possibility that this global economic growth period could also be the superposition of a novel growth phase on the old one, driven by new technologies and using petroleum as the basic source of energy.[30] Surprisingly, Sieferle does not address the main issue of sustainability and relative prices. Environmental historian Joachim Radkau has agreed that the postwar period was "in all likelihood" even "far more important than the Industrial Revolution," although he is not sure in which decade the decisive change really occurred.[31] Christoph Maria Merki has recommended understanding the "1950s syndrome" as a metaphor for a transition period and extending it in three ways: (1) from environmental history to other subdisciplines such as economic and social history; (2) spatially from Western Europe to the entire globe; and (3) finally from the 1950s to the entire period until the present.[32]

Several critics have focused on the name "1950s syndrome" as a point of contention. They have argued that the 1950s was a decade of transition rather than a period of take-off.[33] Frank Uekoetter has gone as far as to claim: "If a prize would be given for the most misleading term in environmental history research 'the 1950s syndrome' would be a leading contender. Designating a single decade in the title distracts from the fact that the thesis refers to the entire postwar period." He further claims that the name "1950s syndrome" hampers the exploration of the early environmental movement in Germany.[34] The designation advanced by the Steffen and coauthors, "The Great Acceleration," would be an alternative to "1950s syndrome."[35] It has the advantage of addressing the longer-term dynamic of the process leading to the loss of sustainability without, however, referring to a syndrome and pointing to the time of the turnover. Perhaps a combination of both labels, such as "the 1950s syndrome or the Great Acceleration," may be the best solution.

However we might designate this phenomenon, we need a better understanding of the obvious but as yet unsettled question: What can explain the surge of global energy use after the late 1950s? This question is best handled by separating it into two further issues: (1) To what extent was the surge of global energy use related to economic growth and/or to population growth? (2) To what extent did the discovery of giant oil fields and the declining prices for fossil fuels in relation to the price of labor encourage the substitution of energy for labor?

The Longest Boom in Human History

The boom lasting from 1950 until 1973 was the longest and the most pronounced in human history, with average growth rates in world per capita GDP of 2.91 percent.[36] During the boom there was a significant degree of convergence in per capita income and productivity, with most regions growing faster than the United States, the largest economy.[37] Per capita growth rates were highest in Western Europe, being "beyond the wildest dreams of even the most optimistic pundit or policy maker in the 1940s. Workers, both skilled and unskilled, and the owners of firms all shared in the benefits."[38] At no time in human history have so many people become affluent within one single generation (figure 7.3).[39]

Not without reason then, these years were named "Wirtschaftswunder" (in Germany), "Les trente glorieuses" (in France, see Jean Fourastié), or the "Golden Age" in the English-speaking world (see Nick Crafts).[40] Sociologist Burkart Lutz has summed up the impression of many contemporaries in the title of his book "The Short Dream of Everlasting Prosperity."[41] After 1973 there was a marked slowdown in the West and in Japan. The best performances came from the economies of East Asia.[42] The shift in the growth pole from the United States in the

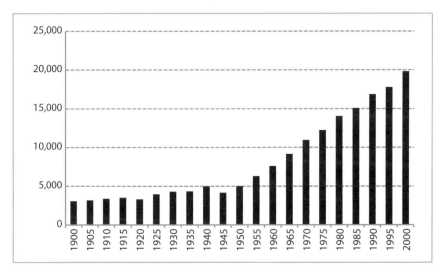

Figure 7.3. Development of GDP in Austria, Germany, and Switzerland, 1913–1996 (1996 $). *Source:* Christian Pfister, "Energiepreis und Umweltbelastung: Zum Stand der Diskussion über das '1950er Syndrom,'" in *Umweltgeschichte: Themen und Perspektiven,* edited by Wolfram Siemann (Munich: C. H. Beck, 2003), 62.

1940s and 1950s, to Europe in the 1960s, and then to East Asia from the 1970s on, is partially depicted in figure 7.3.

Population almost quadrupled between 1900 and 2000, from 1.6 billion to almost 6 billion, and 80 percent of this growth fell within the second half of the century, with rates up to 2.2 percent per year during the 1960s.[43] Obviously, the pace of population growth corresponds closely with that of global energy use, but this correspondence does not necessarily entail causation. Fridolin Krausmann and his colleagues have demonstrated that fossil energy use and greenhouse gas production grew much faster in the northern hemisphere than in the southern hemisphere, whereas population growth was more rapid in the south, which increased biomass extraction and deforestation.[44] Table 7.1 shows that the increase in GDP between 1950 and 1973 in the "rich" countries was slightly below those estimated for the "other" countries, but because the population increase was considerably slower there, the rich nations could increase their per capita GDP far more than the other countries.[45]

Global energy data should similarly be broken down to regional and local levels to allow a more meaningful interpretation (figure 7.4). Our database on energy use for the entire world is incomplete. Evidence for the entire twentieth century only exists for the United States, the Soviet Union, and Western Europe (EU-15). Data for the Middle East, East Asia, and the Pacific (which includes Japan and Australia) as well as for South and Central America are available only from the mid-1960s, and for Africa only from the mid-1980s. The takeoff of energy use began in the United States during the 1940s, in Western Europe and the Soviet Union during the 1950s, and in East Asia and the Pacific area during the 1960s. The figures for the Middle East, Latin America, and Africa are low despite the significant population growth in these parts of the world. Given that the lion's

TABLE 7.1. GROWTH OF POPULATION AND GROSS DOMESTIC PRODUCT IN "RICH" AND "OTHER" COUNTRIES (PURCHASING POWER PARITY $) IN PERCENTAGES (1950 = 100%)

	Population	GDP	p. c. GD	Percentage of global GDP	
	1973	1973	1973	1950	1973
Rich	127	294	231	60	59
Other	163	309	189	40	41

Source: Angus Maddison, *Contours of the World Economy, 1–2030 AD* (Oxford: Oxford University Press 2007), 336–37.

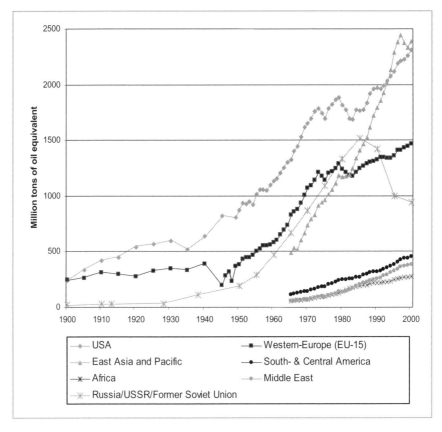

Figure 7.4. Primary energy use in selected regions of the globe, 1900–2000. *Sources:* BP-Amoco, *Statistical Review of World Energy: Full Report Workbook, 2004,* available online at http://www.bp.com (accessed on September 6, 2005); U.S. Energy Information Administration, *Annual Energy Review 2003,* available online at http://www.eia.doe.gov (accessed on September 6, 2005); Brian Mitchell, *European Historical Statistics, 1750–1970* (London: Macmillan Press, 1975); International Energy Agency, *Key World Energy Statistics, 2004,* available online at http://www.iea.org (accessed on September 6, 2005); and Bouda Etemad and Jean Luciani, *World Energy Production, 1800–1985* (Geneva: Droz, 1991).

share of the increase in energy use between 1950 and 1980 is credited to the rich nations, where population growth was moderate, the significance of this factor for the upsurge in energy population growth should not be overestimated. Oil production in the Soviet Union peaked in the 1980s and has since continuously declined, whereas the production of natural gas has expanded.[46]

Why was growth in Western Europe during the Golden Age so fast? This is a complex issue that seems poised to become one of the staple controversies of quantitative economic history.[47] The establishment of an adequate policy framework clearly mattered. The key features of this involved a return to fixed exchange rates under the auspices of the Bretton Woods system—which was not fully achieved before late in 1958—multilateral trade liberalization and economic integration through the establishment of GATT in 1947, the European Economic Community (EEC) in 1958, and the European Free Trade Association (EFTA) in 1960, a rather cautious government intervention into the market economy.[48] In addition, huge investments by the U.S. administration into European economies within the framework of the Marshall Plan, the provision of various forms of welfare by public authorities, educational reform with a focus on higher education, and a kind of "social contract" between labor and capital (see Nick Crafts's work on this subject) were part of this policy framework.[49] It obliged entrepreneurs to make up for inflation by increasing actual wages in proportion to productivity improvements, whereas wage earners backed off from pressing for higher wages.[50] The collective bargaining agreement between the U.S. auto industry and the trade union of the United Automobile Workers yielded the model that subsequently became the standard throughout the United States and also in Western Europe.[51]

Together with the widespread establishment of welfare systems and an improved access to higher education for the middle classes, these collective-bargaining agreements were instrumental in boosting consumption of broad strata of European societies to previously unknown levels.[52] Nick Crafts has emphasized additional key features that shaped the economic environment during the boom: the low initial performance of the economies after World War II, which allowed high growth rates, the extent of technology transfer from the United States, the absence of competition from Asian countries, high rates of investment, and the fact "that energy and other raw materials seemed to be in elastic supply and their prices remained low."[53]

Economic historian Knut Borchardt has remarked that "the industrial world profited extraordinarily during the period of cheap oil."[54] Likewise, Eric Hobsbawm has highlighted this factor, although his reasoning is questionable. All in all, however, the unusually low prices between 1958 and 1973 went generally unnoticed, as Mohssen Massarat has observed, unlike the so-called "oil price shock" in 1973, which was intensively discussed.[55] The fear of a potential scarcity drove prices up, although there was more than enough oil available on global markets. The trade journal *Petroleum Intelligence Weekly* reported in August that "near panic buying by U.S. and European independents, as well as the Japanese" was sending "oil prices sky-rocketing."[56]

The announcement of an energy embargo by OPEC in the context of the Yom Kippur War and the substantial increase of the oil price in late 1973 led to a rude awakening from the "short dream of everlasting prosperity" (in Burkart Lutz's words) in the form of a shock.[57] The usual term of "oil crisis" is inappropriate, because the devaluation of the U.S. dollar and the end of the Bretton Woods system was far more important for the subsequent economic slump in 1974–1975 than the surge in oil prices.[58] According to Knut Borchardt, "the oil price shock was even beneficial for the mood in West Germany, because it allowed the government to blame foreign agents for the recession. The resulting rise in unemployment could therefore be perceived as a kind of natural disaster."[59] In a longer-term perspective, the "oil price shock" needs to be considered as a step toward normalization of relative energy prices.

The Substitution of Energy for Labor

To what extent did the effect of declining prices for fossil fuel in relation to the price of labor encourage the substitution of energy for labor? Economists have given a great deal of thought to the relative price effect and its limitations.[60] In this context it refers to the fact that the relative price of a good not only reflects the familiar model of supply and demand, but is also affected by the price change of other goods. That is, as one good becomes less costly in comparison to others, people will theoretically buy more of this good, which is called "substitution."[61] Hansjörg Siegenthaler, a prominent member of the Regulation School, relates the microeconomic level of individual agency to the macroeconomic level of trade cycles and social change.[62] He has called attention to framing, which means that individuals rely on certain perspectives or "world views" to select, understand, and respond to information.[63] Accordingly, prices per se are meaningless for individual agency unless such changes are interpreted.[64] Interpreting the consequences of relative price effects during the golden age is complicated by the fact that it resulted from two components: namely, a strong rise in real wages and a concomitant moderate price decline for fossil fuel. It is clear that real wages were the focus of contemporary interpretations, whereas declining real fuel prices were hardly noticed.

The second line of argument focuses on the incentives for substituting labor with energy.[65] The unprecedented long boom of real wages, resulting from both the acute shortage of workforce and a rigid wage system, created one set of preconditions for the decline of relative prices for fossil fuel. The other set is related to the enormous available quantity and the dirt-cheap production costs of Middle Eastern oil. When Lee Everett DeGolyer, who directed the U.S. oil explorations in Saudi Arabia during World War II, returned to Washington, D.C., in 1944, he reported "that the proven and probable reserves of the region—Iran, Iraq, Saudi Arabia, Kuwait, Bahrain and Qatar—amounted to about 25 billion barrels." In fact, he suspected that the reserves would be much, much larger. And, indeed, estimates that sounded like lunacy—up to three hundred billion barrels for the region and one hundred billion barrels from Saudi Arabia alone—resulted from his trip. One of the members of his mission told officials in the State Department: "The oil in this region is the greatest single prize in all history."[66] The reserves in the Middle East increased from 28 billion barrels to 367 billion barrels between 1948 and 1972. The Saudi Arabian oil field Ghawar, which is the greatest oil-bearing structure in the world, began to be productive from 1949.[67]

Safaniya, the world's largest offshore oil field, was discovered in 1951, and in 1956 the Saudi Arabian Aramco confirmed the scale of Ghawar and Safaniya.[68] In the same year M. King Hubbert developed the Peak-Oil model, according to which the production rate of a limited resource will follow a roughly symmetrical bell-shaped curve based on the limits of exploitability and market pressures. Hubbert argued that the rents for petroleum would evolve according to the changes of known reserves. As long as reserves are increasing, rents are declining, which was indeed the case until 1972. A cartel of seven multinational petroleum companies (among them Exxon, Shell, Texaco, and British Petroleum) known as "the Majors" or "the Seven Sisters" determined global oil prices until the 1970s.[69] Their profits were enormous compared with the small expenditures for producing the oil and the royalties they had to pay to the producing countries.[70] With a total production cost of around eighty-five cents selling for around $2.50, a barrel of oil was already highly competitive in the late 1940s. During the 1950s production costs plummeted as low as ten cents to twenty cents a barrel.[71]

Not surprisingly, oil production boomed and explorations were rapidly expanded in view of the possibility of such fabulous economic profits.[72] "New oil" increasingly flowed from the mid-1950s to the already expanding production of oil from activated wells. The result was overproduction, despite the enormous boom in demand, accentuated by the fact that "independent" producers increasingly entered the market.[73] The competition among suppliers of crude oil was reinforced by the growing number of refiners independent of "the Big Seven."[74] Both the petroleum companies and the elites of the oil-producing countries strove to pump out the producing fields as quickly as possible. Their mutual interest in doing so, called the "dual system," was rooted in the legal terms of the leasing conditions. On the one hand, the oil companies feared obtaining less favorable conditions after expiration of their contracts. The elites of the oil-producing countries, on the other hand, needed increasing income to persuade their clientele to stick with them.[75] These dual driving forces expanded the overproduction and the already existing pricing pressure.

Far from being just the outflow of market forces, energy prices were at their lowest since the interwar period.[76] Political authorities in Western Europe commissioned a set of studies about the most reliable sources of energy supply in the mid-1950s, the most important of them being the OEEC Hartley Report. Considering several scenarios, such as increasing importation of cheap U.S. coal and promoting nuclear energy, which was still a developing industry, the report finally recommended the import of oil as being the best solution to improve the prosperity of Western Europe.[77] The democratization of consumption became a "basic right during the Cold War, and it was a fundamental argument in the favor of capitalism in the competition with the socialist countries."[78]

Reconstructing Western Europe was the first objective of U.S. policy in the Cold War, and without abundant and cheap Middle Eastern oil this would not have been possible.[79] Governments, hesitating between the strategically and socially desirable maintenance of the domestic coal industries, and the possibility of improving their competitive capability with the United States by opening up their markets to the flood of cheap Middle Eastern petroleum, finally decided in favor of the second option.[80] In fact, the growth rates during the golden age could not have been realized on the basis of coal.[81] The decision of governments to decontrol oil imports was taken just before the crumbling quasi-monopoly of the Big Seven sent prices plummeting from 1957.[82] It needs to be stressed as well that all agents anticipated energy prices to rise both for coal and for oil.[83] None of them expected oil prices to decline in a situation of rising wage pressure and economic pick-up, because such a feature was simply not known from past experience.

After the temporary closing of the Suez Canal during the war in 1956, a new class of very large crude carriers (VLCC) carrying several hundred thousand tons of oil was launched. Transporting a barrel of crude on such a giant was considerably cheaper than on the relatively small vessels on which petroleum had been previously shipped.[84] Pipelines built from the petroleum harbors to the inland refineries contributed to further lowered transport prices.[85] In the end the international price of oil fell sufficiently to pose a competitive threat to domestic fuel industries in the United States.[86] To protect domestic oil producers,

President Eisenhower therefore arranged for a system of mandatory quotas on imports in March 1959, which depressed world market prices even further for the next fourteen years.[87] The combination of these factors led petroleum prices in 1991 dollars to drop 20 percent between 1957 and 1973.[88]

The competitive advantage of oil imports over domestic coal in Western Europe initiated a new conflict during these years, which was mainly reported in the business sector of the media and barely hit the headlines. In its physical properties coal is inferior to oil with regard to energy density, stocking losses, and emissions of CO_2, soot, and SO_2.[89] The change from coal to oil heating mattered, not the least in large smog-ridden conurbations such as London.[90] The higher energy density implies that more units of useful work is provided per unit of thermal input.[91] Besides its physical and environmental advantages, oil is more convenient to use than coal, as it does not need to be scooped into the boiler. Most important, underground-mined European coal was far from being competitive with dirt-cheap Middle Eastern oil. At the same time, U.S. coal mined from open pits became competitive during the 1950s.[92]

Mechanization of underground coal mining was limited to coal-cutting machines, conveyor belts, electric trains, and elevators, leaving much to the hard physical work of miners. As a result, labor made up the lion's share of production cost. Demand rose and fell according to the business cycle involving corresponding fluctuations of wages and energy prices and accordingly the hiring and firing of miners.[93] Early on, European governments were aware of the tightening competitive environment on the energy market through increasing petroleum imports. According to Peter von Allmen's findings, the attempt to create a common market for coal and steel through the foundation of the European Coal and Steel Community in 1951 was primarily intended to improve the competitive ability of domestic coal production.[94]

The shift in the proportionate use of coal and oil had already begun in the early 1950s. Things went haywire in 1957. Miners could hardly be found on the dried-up job market, and coal prices rose substantially.[95] At the same time fuel prices plummeted by 60 percent within three years.[96] The Big Seven weathered the slump by subsidizing part of transport costs from their benefits.[97] As a result, coal shortages ended quickly, being replaced by mounting piles of unwanted coal at pitheads.[98] Governments in the United Kingdom and in West Germany, and to a lesser extent in France, took some measures to protect domestic fuel industries, mostly in view of avoiding or buffering social conflicts. The worst shocks were avoided with the easy availability of new jobs in other branches of the economy and rather generous social compensation plans according to the abundant public revenue.[99]

Within a couple of years the Western European fuel economy was turned upside down. In 1955 coal provided 75 percent and petroleum just 23 percent of total energy use. By 1972 the share of coal had dropped to 22 percent while that of oil had risen to 60 percent. While in 1950 the United States was the only country with a well-developed oil industry, between 1950 and 1973 the industry in the rest of the world grew ninefold at an annual rate of 10 percent.[100] After 1962 crude oil and oil products accounted for more than the half of the global freight volume of international maritime shipping.[101]

The profound change in the structure of the energy sector led to a globalization shock in the field of energy supply that is akin to the globalization shock in grain supply that happened in the early 1880s. The "grain offensive" of the United States and Russia at that time resulted from the opening up of the American West and the Russian plains for commercial grain farming and the concomitant construction of a network of railways and steamboats for shipping the grain to Europe at prices that undersold the production cost of many domestic producers.[102] The glut of cheap overseas grain led to a substantial lowering of grain prices relative to wages, which may have boosted the trade cycle during the two decades preceding World War I. However, the grain offensive undermined Europe's food security with far-reaching consequences in World War I. Likewise, the "petroleum offensive" of the 1950s and 1960s jeopardized Europe's energy security, leaving the region entirely dependent on an outside energy supply and vulnerable to extortion, as the oil price shock in 1973 and more recent events have revealed.[103]

At about the same time energy markets in the socialist countries were reshaped as in the West.[104] From 1955 the Soviets got a surplus of oil, which in part went to the satellites for prices somewhat below the world market level to build up chemical industries.[105] Total Soviet oil exports to all of the COMECON countries grew up to a maximum of ninety-two million tons at the time of "Soviet Peak Oil" in 1981, which fell far short of the import of Western European countries.[106] Not surprisingly, coal remained the most important fuel in Eastern Europe until 1989, which may be one reason why these countries were economically less successful.[107] Coauthors Petra Kuskova, Simone Gingrich, and Fridolin Krausmann have investigated the energetic transition in Czechoslovakia using the approach of material-flow accounting. This country was a center for heavy industry within COMECON. The decades between 1950 and 1981 saw a 400 percent surge in fossil energy consumption, up to a per capita use 2.7 times higher than in neighboring Austria. Environmentally inferior brown coal made up the largest share of this.[108] The situation in neighboring East Germany was hardly better. The resulting enormous air pollution with SO_2 was at the root of "death of the forests" (*Waldsterben*), which typically was rampant in the early 1980s in such regions as the hilly area of Erzgebirge, situated between the CSSR and East Germany.[109] Thus the socialist bloc substantially contributed to the rising levels of greenhouse gases.

In the West the main reason for the substantial decline of the cost of fossil energy in relation to the cost of labor is the effect of two components at the root of the "1950s syndrome": namely, sharply rising real wages and the concomitant availability of dirt-cheap petroleum.[110] Understanding the causes of this "great energy price anomaly," as the feature may be named, is complicated by the fact that the surge in real wages was itself an effect of declining relative prices for fossil fuel to an extent, which still remains to be investigated. The instance of declining prices for fossil fuels during the most unprecedented period of economic growth is a unique feature in human history. It not only contrasts strongly with previous historical experience, but also with the situation in the twenty-first century, where energy prices are again positively correlated with the business cycle.

Among the many consequences of the "great energy price anomaly" are its effects on innovative activity. In 1932, John R. Hicks introduced the theory of "induced innovation," which states that changes in relative factor prices should lead to innovations that reduce the need for the relatively expensive factor.[111] Co-authors Richard G. Newell, Adam B. Jaffe, and Robert N. Stavins found that the nature of the technical change depended on energy prices. High prices tended to encourage energy efficiency improvements. Lower prices led to technological change related to time-saving, implying a substitution to a more energy-intensive production.[112] David Popp has statistically supported this argument by demonstrating for the United States from 1970 to 1993 that both energy prices and the quality of existing knowledge had significant positive effects on energy-efficient innovations.[113]

There is also historical evidence for Hick's theory. For example, energy prices were high in relation to labor before the late 1950s. The forerunner of the steam engine, Thomas Newcomen's steam pump (invented in 1712), was an incredible energy guzzler fed with coal leavings and working with 0.5 percent energy efficiency. Watt's steam engine (developed in 1765) was already 60 percent more energy efficient than the steam pump, whereas the steam turbine built by Parson in 1884 was as much as 90 percent more efficient than Watt's construction.[114] As a result, the cost of steam power in Britain fell from 4.5 pounds in 1760 to 15 pence by 1910.[115] As long as energy was the marginal factor, the quest for economic efficiency likewise improved ecological sustainability per unit of energy services consumed. The reverse can also be true. The sustained decline in the relative energy price during the "great energy price anomaly" might have choked off innovative activity related to energy efficiency. In particular, this refers to the development of sustainable forms of energy use, such as solar, wind, and geothermic energy.[116] Whereas solar homes became increasingly popular in the United States during the late 1940s and early 1950s, "constantly falling fuel prices led to an almost universal lack of interest in solar architecture by the late 1950s."[117]

The Rise of Consumer Society and the Accelerated Loss of Ecological Sustainability

Changes in energy use and greenhouse gas emissions, even far-reaching ones, are not sufficient to justify the designation of a historical epoch. Rather, it must be demonstrated that such changes significantly altered peoples' way of life, their scope of action, their options and aspirations. Such issues as global energy use and relative prices are difficult to bring into line with narratives of people's experience. A cursory reading of the relevant issues, however, may suggest that the change in relative prices in the context of the "1950s syndrome" led to the emergence of "consumer society" in Western Europe.[118] This impression clearly overstrains the explanatory power of the concept. Granted, there is consensus that a Western European "consumer society" did not emerge before the end of the 1950s or the early 1960s—overlooking differences between countries—before "a predominant part of the population substantially consumed beyond its basic needs."[119]

Important components are found in the technological style of Fordist-Taylorist mass production, which unfolded in the United States during the interwar period and in Western Europe from the late 1950s. It allowed for substantial reduction in the cost for durable consumer goods (cars, household appliances, and the like). Few contradict that this changeover grew out of profound changes in the availability of time and money.[120] Likewise, most scholars agree that the consumer society cannot be boiled down to the ownership of new consumer durables like cars and washing machines. For the first time a generation had grown up after World War II, for which austerity and threats to physical integrity in the context of war had no longer been a formative experience.[121] In this context Victoria De Grazia has referred to a "change of a bourgeois mode to a fordist mode of consumption," introducing not only features of mass distribution such as the supermarket, chain retailing, professional advertising, but also more fundamentally eroding the pyramidal class constellation typical of European societies that implied carefully demarcated class-specific systems of "barriers and levels" in consumption.[122] Above all, the consumer society brought a multiplication of options and a diversification of practices.[123] Such changes were cultural rather than income driven and need to be embedded within a history of mentalities that should also consider economic and ecological arguments.

To what extent were the changes in relative prices for labor and energy in the 1950s responsible for the development of the wasteful type of "Western" consumer society? To answer this, we should briefly recall the way of life in Western Europe at the eve of the 1950s. Wages covered basic needs without offering large surpluses for the choice of individual lifestyles. Most of the household budget was spent on food, clothing, shelter, and the raising of children. In war-ridden countries such as Germany, people could survive only by tapping into every possible source of food, especially during the first postwar period.[124] A remark in a student's essay is revealing for the magnitude of the distress. It reads: "My best day was when my brother Friedrich died. Since then I have a coat and shoes and socks and a knit vest."[125] Working-class households still lived on tight burdens in the early 1950s. Coffee remained a Sunday treat throughout the decade.[126]

The situation in France and in the United Kingdom was not much better.[127] Workers had to live in crowded flats within walking, biking, or tramway distance from their place of work. Meeting the needs of a household was a study in rational planning. Food was bought in small "mom-and pop" retail stores.[128] The goods were wrapped in paper bags or sold without any packaging.[129] Fruits and vegetables originated from the surrounding countryside and changed according to season. Such descriptions are not inspired by nostalgia. Rather, they show how the idea of near sustainability was translated into real life. Both men and women, and to some extent even children, particularly in the economy's traditional sector, had to perform hard physical work with no hope of increases in wages.[130] A typical West German housewife worked seventy hours a week in the 1950s—not including the non-wage-earning work that women did for their children, their husbands, and their families.[131] The term "ancien regime ecologique" sums up this situation well.[132]

How sustainable was the "ancien regime ecologique"? The scholars Mathis Wackernagel and William Rees at the University of British Columbia have created a widely recognized comprehensive ecological accounting system for assessing the environmental impact of human activities known as an "ecological footprint."[133] The ecological footprint measures the amount of biologically productive land and sea area human activity requires to produce the resources it consumes and absorbs the waste it generates, and compares this measurement to how much biologically productive land and sea area are available. Ecological overshoot occurs when a population's demand on an ecosystem exceeds the capacity of that ecosystem to regenerate the resources it consumes and to absorb its wastes. Most "northern" countries had an overshoot of more than four global hectares (gha) per person in 2005, with the United States leading the list. Historical figures are available back to 1961.[134]

The example of the Netherlands is representative for Western Europe (figure 7.5).[135] From 1961 to 1973—that is, during the period of most rapid divergence between real wages and energy-prices—the ecological footprint in that country shows a continuous strong rise. Since 1973, however, it has fluctuated on the same level. To what extent was the effect of relative prices related to human decision making? This issue needs to be broken down, with production and consumption being the two major fields of economic activity. Regarding consumption, we need to distinguish between buying commodities or services and making

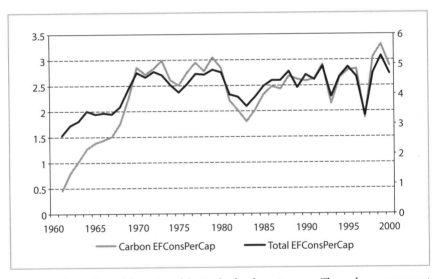

Figure 7.5. The ecological footprint of the Netherlands, 1961–2000. The carbon component of the "ecological footprint" translates the amount of carbon dioxide into the amount of productive land and sea area required to sequester carbon dioxide emissions. *Source:* Brad Ewing, Steven Goldfinger, and Mathis Wackemagel, *The Ecological Footprint Atlas* (Oakland, Calif.: Global Footprint Network, 2008), 36. Data kindly provided by Global Footprint Network, National Footprint Accounts, 2008, available online at http://www.footprintnetwork.org. For more information about the footprint methodology and calculation standards, see http://www.footprintnetwork.org.

longer-term investments. Individual preferences prevail in buying clothes, furniture, consumer electronics, and services, whereas economic arguments might have been more significant in making longer-term investments, such as choosing the place of residence or the size of the family car.

The case of production of goods and services, including packaging and distribution, is more obvious.[136] Firms and governments make decisions according to the laws of the market economy. In a period of galloping real wages, a myriad of decisions made at the level of the individual firm were targeted at saving labor, or time, which is much the same thing. Such measures were justified as "rationalization," "savings of costs," "improving the quality of location," and so forth. Disregarding improvements in organization, most measures involved additional inputs of fossil energy without explicitly addressing this factor, because it was not cost relevant. The specific logic of such decisions widely differs according to economic sector. Agriculture is possibly the best suited case for demonstrating the effect of changes in relative prices, because the productivity increase in this sector known as the "Green Revolution" was by far the most outstanding.[137]

The "Green Revolution"

"Modernizing" agriculture in Western Europe in terms of increasing yields per hectare and raising labor productivity was a long-lasting process going back to the Middle Ages.[138] Paul Bairoch has distinguished three phases. During the first, organic phase, agricultural production was essentially "solar," depending on recycling inputs grown on the farm and a substantial input of labor. The second, mechanical phase from the mid- to late nineteenth century involved the application of more efficient tools, horse-drawn machinery, and a limited amount of inputs from outside the farm. The third phase, that of "high-input agriculture" beginning in the 1950s, marks a significant breakaway from traditional agriculture, using renewable solar energy to produce a surplus of consumable biomass based on soil, plants, and photosynthesis.[139] High-input agriculture involves a set of heavy motorized labor-saving machinery introduced in combination with such land-saving techniques as high-yield varieties, depending on massive inputs of fertilizer and pesticides.[140] Rolf Peter Sieferle and his coauthors have demonstrated that cultivated areas ceased to be net producers of biomass for human and animal consumption through this transition, because the input of fossil energy exceeded the energy content of the biomass harvested.[141] The Green Revolution was not limited to Western Europe; it also became a major feature in Asia and Latin America.[142]

Hermann Priebe has considered the development of European agriculture before World War II as being a telling example of "organic development": "A centuries-old epoch, during which renewable resources were carefully managed, ended in the post war period."[143] Indeed, before the mid-1950s most of the farmers still worked with horse-drawn machinery. They relied on their own fodder, and most of the manure was produced directly on the farm. Weeds were mechanically suppressed and bugs were picked out by hand.[144] For the Austrian peasant the Middle Ages lasted until the mid-twentieth century.[145]

After the war the immediate concern throughout Western Europe was raising agricultural production. Besides food shortages, there was a general need to maintain a good balance of trade by keeping imports as low as possible. The recovery, assisted by Marshall Plan funds, was extremely rapid. Labor productivity increased faster in agriculture than in other sectors. It tripled in France and Denmark between 1960 and 1980. Fridolin Krausmann has summarized

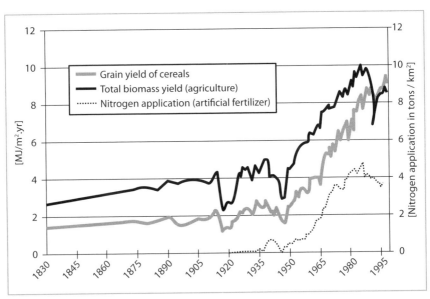

Figure 7.6. Grain yield of cereals, total biomass yield agriculture, and total nitrogen application (artificial fertilizer) in Austria, within the borders since 1830. *Source:* Fridolin Krausmann and Helmut Haberl, "The Process of Industrialization from the Perspective of Energetic Metabolism: Socioeconomics Energy Flows in Austria, 1830–1995," *Ecological Economics* 41, no. 2 (2002): 177–201.

the course of agricultural development in Austria from 1830 to 2000, assessing biomass yields, grain yields, and the input of artificial nitrogen fertilizer in a comprehensive graph (figure 7.6). This 170-year-long period is divided into two phases of unequal growth. Before 1950 annual biomass yields grew at an average of 1 percent. Subsequently, annual growth rates of biomass production and yields jumped to 4.5 percent, which, among other causes, was the effect of a massive input of artificial nitrogen fertilizer.[146]

Considering the rate of increase in productivity and input of artificial fossil fuels, the transition from traditional to high-input agriculture certainly deserves the term "revolution" in the intrinsic sense of a rapid, deep-reaching, and irreversible process.[147] Eric Hobsbawm has pointed to the fact that the Green Revolution led European peasants to vanish and that this outcome, involving a significant loss of culture and traditional practices, is irreversible.[148] That said, it needs to be stressed that in Asia and Latin America the Green Revolution was crucial for feeding the rapidly expanding populations.[149] Indeed, after 1950 real agricultural prices remained stable or even decreased in a survey of 130 countries.[150]

To assess the extent to which Europe's Green Revolution was promoted by changing relative prices for labor and (fossil) energy, prices of wage labor, capital, and energy have been collected from primary sources in Switzerland, for which reliable data exists.[151] At first glance Switzerland would seem not to be an appropriate candidate for a case study of agricultural economics in Western Europe, since it is outside the European Union. The EU has directed the agricultural policy in the Community's member countries since the 1950s. However, the principles of Swiss agricultural policy were very close to those of the EEC, except that tariff barriers were even higher. The incentive structure set by agricultural administrations both in the EEC and in Switzerland encouraged the production of nearly unlimited quantities of milk, grain, potatoes, and so on, because target prices were paid regardless of the amounts produced.[152]

Considering the evolution of foreign wage labor, machinery, diesel, and artificial nitrogen fertilizer (figure 7.7), the massive changes in relative prices between 1948 and 1973 are striking. The price of wage labor grew almost sixfold, that of agricultural engines scarcely doubled, whereas fuel and fertilizer prices remained almost constant. It needs to be stressed that energy is the main ingredient used in nitrogen fertilizer (ammonia) production, accounting for more than 80 percent of the production costs.[153] The significance of declining relative energy prices needs to be considered within an economic environment shaped by technological push factors and political agents who, to keep some farmers in

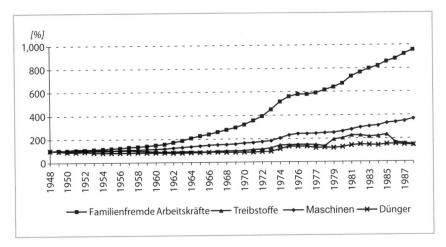

Figure 7.7. Relative prices of foreign wage labor, machinery, fuel, and nitrogen fertilizer in Swiss agriculture, 1948–1988 (1948 = 100). *Source:* Christian Pfister, "Die Umwälzung der Landwirtschaft im Tieferen Mittelland seit den 1950er Jahren und der Untergang des Bauerntums," in *Die Bauernhäuser des Kantons,* edited by Heinrich Christoph Affolter, vol. 3 (Bern: In press).

business, strongly promoted the transition to high-input agriculture. Overmechanization at the expense of wage-labor was for larger farmers a matter of survival. Likewise, considering high fixed costs and payment of interests for machinery and low costs for fuel and fertilizer, expanding production and pushing yields to a maximum became profitable (figure 7.7). Although the availability of cheap fossil energy was of primary importance for the transition, classical studies of agricultural economics did not address this issue at all.[154]

The lion's share of the productivity gains benefitted the consumers. Analyzing household consumption patterns in France, West Germany, and the United Kingdom, Sabine Haustein has concluded that expenses for food declined most substantially between 1945 and 1970. The extra purchasing power was chiefly spent for "transportation"—that is, for driving a car in the context of leisure activities.[155] This further boosted fossil energy use and promoted economic growth in terms of a feedback effect. The environmental consequences of high-input agriculture—such as loss of biodiversity and landscape amenities, nitrification of groundwater due to overfertilization, soil erosion, and compression—were summarized by Antoinette Mannion.[156]

Mass Motorization and Mobility

The automobile was and still is undoubtedly the key product of modern consumer society. Together with the family home in suburbia, it provided the basis upon which the household existed.[157] The car as a commodity made its breakthrough in Western Europe between the late 1950s and the early 1960s. Peter Borscheid has spoken about an "epoch-making divide in the history of mobility."[158] Indeed, when the writer Philip Bagwell came to live in a north London suburb in 1953, "the motor car was still not completely dominant. Private car ownership was limited to the occupance of one house in every five. Milk was delivered by means of a horsedrawn car. . . . Ten years later horse transport had disappeared from the neighbourhood and car ownership had extended to three quarters of the households."[159]

Mass motorization in Germany was promoted by declining selling costs for cars—for example, by 20 percent between 1948 and 1957 for VW's Beetle—and by lowering the annual tax on motor vehicles in 1955 by 20 percent.[160] Looking at pictures of the cars of the 1950s reveals that small cars were produced in response to the high gas prices at that time. The Morris's Minor, Citroën's legendary Deux Chevaux, Fiat's Seicento, and VW's Beetle dominated the streets, not to mention the "cabin scooters"—such fancy hybrids as the Goggomobil. In 1950 the railways still carried the lion's share of passengers, but they were under a triple threat from private cars, motor buses, and airplanes.

According to Christoph M. Merki, we would be ill advised in interpreting the "jolty triumphal procession of the automobile" solely as an effect of relative prices. It should rather be understood as a holistic process involving economic, sociological, technological, and cultural perspectives.[161] As people became more affluent, they appreciated the advantages of a motor car with its door-to-door-transport, convenience, privacy, and, once the fixed costs were paid, low unit cost for family transport.[162] One of the main motivations for buying a car was the expectation of saving time. This was only possible if the network of roads was adequately developed. The building of expressways was a key promoter of mass motorization. Leaving Italy and Germany aside, the construction of expressways in most European countries was planned and decided in the 1950s. For Gijs Mom the emergence of the limited-access highway was a turning point in the history of mobility.[163] "Autobahn" networks were imposed on the previous system that had been designed on different principles.[164]

During the 1950s the International Federation of Roads (IFR) lobbied for the diffusion of American road-building techniques and promoted the formation of an "elite corps of highly trained road administrators over ministers and directors" as well as five hundred engineers.[165] Mom has argued that the building of such networks by states or state-sponsored entities "was not a response to actual demand (in terms of the number of cars used for long range mobility), but rather to expectations that these vehicles would be more widely used in the future."[166] It was fantasy, inspired by the existing railway network, and eventually road traffic came to resemble railroad traffic, in that roads caused traffic to become long-range rather than peri-urban or regional.[167] At the same time such views reflected the public's desires, as designed motor roads were a visible sign of progress intended to reduce accidents and to speed up traffic flows.[168] In Switzerland 80 percent of male voters approved a corresponding amendment to the constitution in 1958.[169] In 1963 construction was barely under way in most European countries, but by the end of the 1980s the European highway network had increased to forty thousand kilometers.[170]

The price of gas was politically determined from the very beginning of the automobile age.[171] The cost of expressway construction was in most countries covered by an extra tax on gasoline. However, with rising wages and declining gas prices, this tax became less and less a burden, which in fact led to an "auto-financing" of expressways in a double sense of the word.[172] What about relative prices? Data about wages, the prices for gas, electricity, and brown bread were collected for Switzerland between 1950 and 1990 (figure 7.8).

The cost of a liter of gasoline in 1950 was more than that of a kilogram of brown bread. A skilled worker's hourly wage was worth 4.5 liters of gasoline at that time. Forty years later, it was equivalent to more than 20 liters of gasoline. In relation to wages, gasoline had become five times cheaper during this period, including a surtax for motorway building and maintenance.[173] How deeply this experience altered people's attitudes was revealed in 1973 when a "right to gas" was claimed in the Swiss press as a clear analogy to the traditional "right to bread."[174]

That gas prices were hardly an issue up to 1973 is precisely the key point from an ecological point of view. Once the fixed costs were paid, the mileage covered per year hardly made a difference. Low mileage was an omnipresent advertising argument until the late 1950s.[175] Subsequently, cars were increasingly being used for walking or biking distances that had been unthinkable before 1960. As a matter of fact, the proportion of national income spent on travel and transport alone in Western European countries roughly doubled between 1946 and the late

1980s.[176] The significance of the relative price effect prevailing in the 1950s may be demonstrated by assuming a rise of the gas price in proportion to real wages until 1990, which would correspond to as much as 6.60 Swiss francs of 2008 or 4.30 Euro per liter.

Nobody had anticipated the rise to predominance of road haulage in the 1950s. The growth in truck size and the greater power of engines spurred by the new expressways allowed both larger loads and greater speed. The great economic advantage of road transport was not so much cost, although that fell in real terms, as its predictability, flexibility, and speed. Using trucks meant a journey with an "unbroken seam," that is only one loading and unloading, whereas railway travel usually doubles that number.[177] Again the significance of the diesel prices needs to be considered from an environmental perspective. As the bulk of the costs were fixed, the mileage did not matter. As a result, long-distance road haulage became profitable, which had never been the case before. The availability of cheap and quick transport on the road and also in the air (figure 7.9) led to a

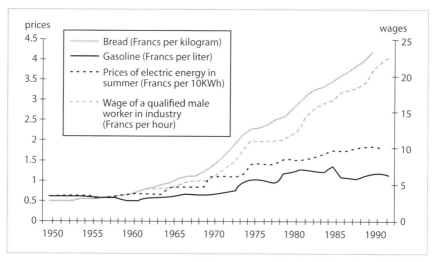

Figure 7.8. Hourly wages in industry compared with prices of brown bread, electric energy, and gasoline in Switzerland, 1950–1990. *Source:* Christian Pfister, "The 'Syndrome of the 1950s' in Switzerland: Cheap Energy, Mass Consumption, and the Environment," in *Getting and Spending: European and American Consumer Societies in the Twentieth Century*, edited by Susan Strasser, Charles McGovern, and Matthias Judt (New York: Cambridge University Press, 1998), 359–78.

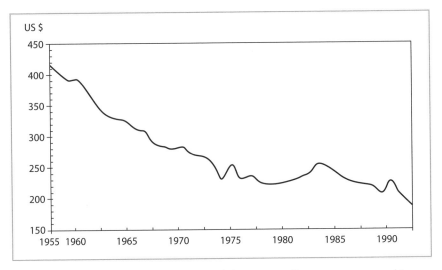

Figure 7.9. Average European fare for a 1000-kilometer trip (in constant 1993$ and to 1993 prices). *Source:* Hans-Liudger Dienel and Peter Lyth, *Flying the Flag: European Commercial Air Transport since 1945* (Basingstoke: Macmillan Press, 1998), 11.

relocation of firms and residential areas akin to the rearrangement of the spatial matrix of human activity through nineteenth-century railways. At the same time it led to a further quantum leap in globalization.

Air transport has seen the most impressive growth rates of any mode of transport since the 1950s. In Belgium, France, Spain, and the United Kingdom, for which data are available, kilometers flown by passengers grew 3,900 percent between 1950 and 1970. The surge in air transport can be explained by factors on both the demand and supply side.[178] It is said that World War I taught people to fly, and World War II gave them the aircraft to do it economically.[179] Most of the first generation of commercial jets, following the pioneering Comet built by De Havilland in 1952—such as Aeroflot's TU-104 in 1956, Boeing's B707 in 1958, McDonnell-Douglas's DC-8 in 1958, and Sud Aviation's Caravelle in 1959—went into service in the late 1950s.[180] The advantage of jets was their superior speed to piston engines, which considerably extended the distance that could be covered per unit of time. Jets were also capable of development in terms of power output, so that even larger planes could be built with lowered unit costs. The globalization of industries and institutions promoted the airplane as a business tool, whereas the increase in real wages and the growing availability of annual vacations brought long-distance holiday travel within the reach of the middle class.[181]

The price history of average airfares since 1950 (figure 7.9) depicts the two components of the "1950s syndrome" well. The surge in real wages and the decline in real fuel prices led airfares to drop substantially before 1973. Subsequently, the curve mainly depicts the effect of fuel prices, particularly the two "oil crises" and the negative oil price shock after 1986.[182] Railways suffered from the competition of both road and air transport. In most countries the dense rail network built up from the mid-nineteenth century until 1914 was partly dismantled in the 1960s and 1970s.[183]

DURING THE PAST two centuries humans have become geological agents and their continuous unbridled use of fossil fuel threatens to "burn" the planet. According to the historian Dipesh Chakrabarty, this fact has resolved the fundamental assumption of Western (and now universal) political thought to separate human history from natural history. Explaining this catastrophe in the moment of the danger that is climate change now calls for a new universal history of humans.[184] In particular, when and why the climate problem that plagues the world today has gotten to this state of calamitous emergency. The economist Nicholas Stern made it clear in his famous analysis that the unbridled rise of CO_2 concentration in the atmosphere is an expression of "the greatest and widest-ranging market failure ever seen" without answering the issue of "when" and "why."[185] This task is left to the historians. Most of them point to industrialization as being the fall of humankind, and this point of view is echoed by scientists. However, the work devoted so far to the "1950s syndrome" and the "Great Acceleration" has made it clear that the greenhouse problem in its present urgency mainly results from the development within the second half of the twentieth century.

A closer examination of the process leads to the conclusion that it mainly originated in the so-called golden age (between the late 1950s and the "oil price shock" of 1973). Two sets of drivers need to be highlighted: namely, a boom in real wages, the most pronounced in human history, which was to a considerable extent a by-product of the Cold War, and a concomitant decline in real oil prices, which was mainly due to the inconceivable enormity of the Middle East oil bonanza. Just the world's largest oilfield, Ghawar in Saudi Arabia, until 2005 yielded sixty billion barrels of crude corresponding to 9.5 cubic kilometers.[186] While it took hundreds of millions of years for its formation, humans (mostly Western societies) will have returned its concentrated organic carbon to the atmosphere and oceans within two or three generations. In 1949 the petroleum geologist M. King Hubbert made public "that the fossil fuel era would be of very short duration."[187] He was right in predicting that the United States would reach Peak Oil in 1970.[188]

The "great energy price anomaly" is the main reason why the Northern type of industrial society became so wasteful. The anomaly lasted long enough to effect structural changes, such as relocating firms and residential areas in a manner predicated on car transportation, restructuring the job market, and intensifying international division of labor. It fundamentally altered people's way of life, moving perceptions and dreams away from the focus on work and saving, toward leisure and spending. In historical perspective current efforts of emerging global governance to substantially lower CO_2 output under the threat of the greenhouse problem might be understood as attempts to get rid of unsustainable practices inherited from the "1950s syndrome" and the resulting "Great Acceleration."

8

Modern Environmentalism

JENS IVO ENGELS

      Divisions into periods are meant to give history a structure and at the same time to fix a certain interpretation of this history. This is all the more true with respect to contemporary history and especially to very recent developments. The history of the past century has seemingly been structured by more important events and "turning points" than former centuries; the past decades seem to have had more significant outcomes than former decades, and so on. On the one hand, recent developments have indeed more direct impact on our time, so that we need a closer look to explain the present state of the world. On the other hand, the perceived acceleration of "eras" and "turning points" is risky. Growing distance in time will inevitably reassess current interpretations. What seems today a crucial historical redirection may be viewed as a secondary incident in ten or twenty years. I therefore take a critical look at one of the most popular turning points of twentieth-century environmental history: the ecological turn (around 1970). Often, the development of the modern environmental movement, environmental consciousness, and environmental politics on the national and international level is depicted as a fundamental redirection of humankind's attitude to nature and the environment, putting to an end the hegemony of the modern exploiting *Homo faber*, who had dominated Western attitudes since early modern times.[1]

    This interpretation is not the least due to the fact that environmental historiography itself is a product of the ecological turn. The first generation of environmental historians was inspired by the politics of the environmental movement and its alarmist attitude concerning the state of the earth.² Environmental historical research had to contribute to the growing realization of the environmental question and had to explain why modern (industrial) societies turned the "wrong" direction—supposing that better ways had to be followed in the future. Early environmental history was therefore by definition part of a fundamental "turning point." So every reflection upon the ecological era is at the same time a reflection upon the history of environmental historiography. Historicizing the ecological era implies historicizing the environmental history project itself. Although the development of political ecology and environmental protection policy is an international phenomenon, I concentrate on the case of (West) Germany. It often has been cited as an example not only of advanced modernity, but above all as one of the strongholds of environmentalism. So it seems to me a good example to test the ecological turning point hypothesis: if the German case encourages skepticism about it, this is all the more true with respect to other countries.

However, the ecological era is not only a category put forward by environmental historians. Fortunately, "general" history has taken some interest in the time around 1970 as a turning point, referring to a multitude of different factors. For some years now, German economic, social, and cultural historians have been advocating a new division into periods for the twentieth century. They refer to the so-called era of "classical modernity," which includes the time between 1880 and 1970–1980. This idea has been inspired by a multitude of empirical studies that suggest that the famous caesurae of 1914–1918 and 1945 did not affect long-term developments of economic trends, social structures, and cultural norms. Looking for a valid conceptual framework, they have adopted the scheme of modernity and "reflexive" or "second" modernity, advocated by the German sociologist Ulrich Beck and his British colleague Anthony Giddens.³

Around 1970, they have argued, second modernity replaced the first one. Reflexive modernity is no longer dominated by the old conflicts about economical distribution and the modernization of the remains of the traditional world. Instead, it allegedly is characterized by the fading of class society and by "postmaterial" value systems. Whereas social problems in classical modernity had been solved by the distribution of ever-growing wealth, economic growth has come to an end. Beck has coined the concept of the "risk society," which is dominated by new kinds of universal risks, threatening all social strata without any difference. Instead of distributing wealth, the societies of second modernity have to protect themselves from risks, most of them created by the secondary effects of modern science, technics, and lifestyle—one of the most important being global environmental destruction.

At the same time the new social movements (including the ecological movement) are depicted as agents of innovative solutions, because they allegedly do not suffer from the deficits of legitimacy that dominate the realities of mainstream politics in Western democracies. According to Beck, these deficits are mainly caused by the fact that scientific expertise has replaced the "political" debate about central issues of the risk society. For the moment these assumptions are merely hypothetical, because historical research on the 1970s and 1980s is just starting. It is nonetheless fruitful for environmental historians to participate in the debate to make an important contribution to "general" history.[4]

What are possible criteria to determine whether the ecological turn constitutes a major turning point? I focus on three dimensions: (1) the state of the environment, (2) environmental consciousness and politics, and (3) environmental protection practice. However, these dimensions are interrelated. Concerning the state of the environment, there are two possible directions of change. In the eyes of many contemporaries, humankind had reached a critical state in the 1970s. In the eyes of the authors of the "Report to the Club of Rome" in 1972 and many contemporary commentators, for instance, humans had the last opportunity to adjust the regime of resource use to avoid collapse.[5] So we have to answer the question if the state of the environment degraded in a way that justifies the rather gloomy visions of Beck. Or perhaps, just the opposite was true: Did the ecological awakening in nearly all Western societies spur a fundamental change in the use of resources, and did it cause a considerable reduction of environmental alteration? Or do we have to define the ecological era on the level of politics rather than on the effects of (environmental) policy: Did environmental consciousness and new environmental protection strategies become central preoccupations of our political systems since the 1970s? Finally, we have to decide if all or any of these criteria underwent a *significant* change in the 1970s, or if their development is better characterized by the term "evolution".

The State of the Environment

With respect to environmental degradation and pollution, only very weak arguments support the thesis of an epoch-making change at around 1970. In general, it is one of the most difficult, if not impossible, tasks of environmental history to assess the state of the environment in a general manner. The diversity of environmental factors includes biodiversity, pollution of the air, water, soil depletion, noise production, heavy-metal-containing food, and so on. Their impact on human life differs from region to region, and their interplay is difficult to conceive. Nevertheless, many of these factors worsen the quality of life of many persons and even threaten their health. Polluting industries and consumer behavior supported the improvement of the quality of life. There is no way to offset the positive outcomes of modern industrial society against its negative dimensions. The only general indicator that can be considered as "hard" is life expectancy. This is thoroughly anthropocentric—and not only related to pollution and the environment, but accessorily to the result of medical treatments, wealth, working conditions, recreation facilities, food situations, the impact of wars and political unrest, and so on. There is simply no way to isolate the state of the environment from other factors influencing the life conditions in modern societies. In fact, life expectancy in Germany (as in other Western countries) rose steadily in the course of the twentieth century and continues to do so. From this point of view, there is little evidence supporting the vision of a highly dangerous "risk society" we allegedly live in since the 1970s.[6] Concerning the world level, so-called ecological optimists like Bjørn Lomborg have maintained that life expectancy and quality of life (with respect to environmental factors) did improve since the early 1970s.[7]

Nevertheless, we cannot say that environmental destruction has declined during the past decades. In fact, contemporary societies face many important degradations. In Europe pollution is doubtless one of the most important problems. Concerning periodization, most environmental historians agree with Swiss historian Christian Pfister and his concept of the so-called "1950s syndrome."[8] According to him, contemporary industrial pollution is more and more produced not by production processes but by consumers. In his research the 1970s do not represent a time of fundamental change. Indeed, Pfister has shown the importance of the 1950s, highlighting the effects of decreasing costs for energy on consumer culture and the (decentralized) production of pollution by cars, washing machines, and other household appliances. There can be no doubt that many processes causing polluting effects linked to consumption (like motorization and suburbanization) accelerated during the 1960s and the 1970s. But the environmental deterioration caused by these processes reached its new *quality* just one or two decades before.[9]

However, there are some important studies underlining the success story of environmental protection measures taken some decades earlier. In West Germany, for example, some leading data of water and air pollution have been reduced since the 1960s and 1970s. Let us take the example of air quality. Between the mid-1960s and 1980, West German nitrogen oxide emissions increased by 60 percent, whereas emissions of organic compounds grew by 10 percent. During the same period the sulfur dioxide load of the air remained at the same level (with the exception of a couple of years at the end of the 1960s, when it rose temporarily). Dust was reduced by 50 percent, and the most important success of West German pollution control policy in this era was the reduction of carbon monoxide, which was lowered by 70 percent. The energy consumption pertinent to aerial emissions increased by 40 percent.[10] The loading level of surface water underwent significant changes too. At around 1970 the chloride concentration in the river Rhine reached its peak, so that biodiversity had been reduced dramatically. By 1980, however, water quality had improved on a level comparable with the early 1960s. Most of these problems were caused by industrial emissions, but private wastewater had growing effects. Administrative measures have been taken since the late 1950s. Between 1963 and 1979 the percentage of private households connected to biological sewage treatment plants rose from twenty-five to seventy.[11]

These findings have to be discussed in a broader time frame. On the one hand, the pollution load has increased in Western societies since the beginning of industrialization, because industrialization has enabled humans to mobilize energy and resources in an unprecedented way.[12] On the other hand, industrialized societies have also developed a wide range of pollution control instruments that mitigate pollution. These measures have not stopped pollution or humans' growing manipulation of the environment. Nevertheless, they have lowered the most important factors injurious to human health.[13] The general trend since the late nineteenth century was not stopped, nor was it significantly altered by and around 1970. Therefore the state of the environment does not support the idea of an environmental turning point during that decade. Moreover, the history of environmental pollution in industrialized societies is such a complex phenomenon (combining degradation *and* improvement), that it seems difficult to deduce any division into short time periods from it at any rate.

Environmental Consciousness and Politics

The most pertinent arguments supporting the idea of an ecological turn around 1970 are related to modern environmentalism as a social and political movement in Western societies. The ecological turn is thus a phenomenon concerning the household of society rather than the household of nature. Although nature protection has a long tradition reaching back to the turn of the twentieth century, in the late 1960s (North America) and the early 1970s (Europe) a new awareness of nature arose. Within a couple of years the deterioration of natural living conditions gained the status of a key political issue. National governments declared environmental protection policy one of the most important challenges to them. Public opinion was also deeply transformed by the idea of environmental destruction. In the several years after the publication of the "Report to the Club of Rome" in 1972, many debates on the environment in Western countries were dominated by the idea of an imminent ecological crisis.[14] A considerable number of political leaders, scientists, and journalists shared a spectacular doomsday vision: the survival of humankind was in danger. There seemed no way to escape from a global environmental catastrophe if humans did not change their modes of production and consumption.

The environmental historian Patrick Kupper has called this the "1970s diagnosis." He underlines that there was no causal link between an alleged deterioration of the environmental conditions and the social construction of the ecological crisis. Indeed, the diagnosis can be interpreted as a social and cultural phenomenon. Following Kupper, the ecological catastrophe was a kind of symbolic discourse articulating experiences of social change, including the gender and women's rights question, the growing problems of the Third World, and the growing "unrest" of the younger generation.[15] Modern environmentalism is characterized by at least four important elements:

- The notion of "the environment" served as an umbrella allowing discussions of various and differing problems that had been treated separately before to be integrated. It enabled a holistic view of and a synthetic approach to such questions as industrial pollution, landscape degradation, health issues, protection of animals, traffic, resource scarcity, the consequences of private consumption, the dangers caused by nuclear power, and so on.
- The environmental protection discourse adopted a global perspective, because nature's degradation transgressed national boundaries and even continents; humankind *as a whole* seemed to be endangered.
- The discussion about the state of the environment became essentially scientific, whereas traditional nature conservation had been presented as a stronghold of antimodern cultural critics. Environmental protection often relied on the new science of cybernetics, or at least on the conviction that nature was a complex system and that the relationship between humans and the environment had to be explained by science. The "Report to the Club of Rome," for instance, sounded plausible enough, because it was based on the computer-generated models of futurologists. Their warning seemed all the more pertinent as they had

advocated, during the preceding years, a thoroughly optimistic vision of the decades to come.[16]
- During the 1970s a massive politicization of the environment took place, followed by an unprecedented success in mass mobilization. In previous decades environmental destruction and nature protection often had been articulated as nonpolitical issues. During the postwar period, for example, the public and the leaders of conservation societies had adopted the idea that environmental problems could and should be solved by consensus or outside the political arena, as being a matter left to technical experts. Now the environmental question was framed as a public and political concern to be debated by all. The environmental movement became one of the most important branches of the "new social movements," restructuring political commitment in Western societies since the 1970s.[17]

Environmental discourse was founded on the basis of science (also called "the knowledge society"), of politicization, and a growing awareness of the global dimensions of human life. The ecological turn combined elements that reduced the complexity of reality on the one hand ("the environment" as an umbrella), but stressed the complex scientific nature of environmental problems on the other hand. The combination of complexity with rather simple slogans was perhaps the most important innovation of the environmental protection concept.

Let us now look at the *scientific nature* of publicly debated environmentalism. The warnings were not entirely new, as medical or technical experts had been concerned with pollution since the early industrial age.[18] Nevertheless, the umbrella of modern environmentalism restructured the knowledge about environmental deterioration and allowed, for the first time, general assessments of the state of the environment on a world level. Moreover, the biological subdiscipline of ecology, which had quickly developed since World War II, claimed to give a precise account of the state of the environment. Ecologists asserted that their knowledge about the interaction of species and their environment enabled them to define solutions for human-made environmental destruction and to show ways that would guarantee the survival of humankind. Temporarily, the advocates of ecology claimed even that it represented a key science for social problems. For several years there was a kind of "ecologymania"—expecting clear solutions developed by rigorous scientists. It was one of the most important political results of the scientific framing of environmentalism that the protection of nature no longer was a concern of isolated cultural critics (as it had seemed to be before). It finally belonged indisputably to the realm of "hard facts."[19]

The political movements quickly adopted the "ecological" formula. As more and more grassroots citizens' initiatives addressed local (and global) environmental problems, they relied not only on scientific expertise. They also claimed that their mode of political commitment itself was "ecological." Environmental activism was called, at the end of the 1970s, "political ecology." The whole complex of "soft" and nonviolent development schemes for both industry and social relations, grassroots politics, and environmentally friendly consumption was summarized by political ecology.[20] In 1980 the newly founded West German Green party published its basic political program. In the program's first paragraph the authors claimed that all their (ecological) demands were based on the objective findings of natural science and only obeyed the "laws of nature." From this perspective good politics seemed to be reduced to the correct interpretation of nature.[21]

The Greens used the argument of the "force of circumstances" (*Sachzwang* in German). This was a paradox because the Greens and other new social movements had started their political activities in the 1970s by denying any argument based on the force of circumstances. Indeed, planning experts, economists, and technicians had argued that economic growth, nuclear power, the construction of roads, and the implantation of new industrial plants simply were economic and social necessities. However, "political ecology" proved to be a forceful symbol of environmental and social-political reform during the 1970s and 1980s.

The global dimension of environmentalism, however, was not entirely new. Since the 1940s, American "new conservationism" had highlighted worldwide resource scarcity, the dangers of population growth, and the interrelatedness of environmental destruction all over the planet.[22] Pollution control experts had met on several occasions at international conferences.[23] The same is true with respect to animal protection. The tradition of international conventions on the protection of species reaches back to the first decades of the nature protection movement.[24] In addition, the famous pre-ecological popular doomsday literature of the 1940s and 1950s never restricted themselves to national issues. The best-selling books by authors such as the conservationist Fairfield Osborn and the ornithologist William Vogt systematically stressed international phenomena.[25]

Nevertheless, the global dimension of environmental destruction gained new political significance because of the changing political contexts. On the one hand, conservative neo-Malthusian publicists like the German-Austrian biologist Konrad Lorenz combined the specter of overpopulation with the old idea of cultural and biological decay—an argument that had been put forward for decades by eugenicists and cultural critics.[26] On the other hand, the student movement and its protest against the Vietnam War had inspired, since the late 1960s, a growing awareness of postcolonial inequalities between Western nations and the countries of the Third World. In this context environmental destruction was interpreted as a symbol of Western exploitation from which the poor populations suffered.[27]

One of the other important characteristics of the ecological turn was politicization. Many governments put environmental protection on their agenda—this was the case, for instance, in the United States, France, and West Germany. Political and cultural elites were convinced that environmental awareness of the masses had to be encouraged. Thanks to important media coverage, governmental initiative, and processes in civil society, the notion of the environment and its destruction spread quickly. Since the ecological turn, environmental issues have alternated between conflict and consensus. During the early years governments hoped they could rely on unanimous consensus. Events like Earth Day and the European conservation year (in 1970) had been conceptualized as harmonious events, allowing the political elites to focus on an undebated area of activity. But in the first half of the 1970s, environmental issues increasingly became occasions to articulate social, cultural, and political unrest. It turned out to be a phenomenon of protest rather than official policy. The reasons for this radical turnaround have been discussed elsewhere.[28]

Governmental commitment was doubled and even outdone by private initiative. In most Western countries ecological subcultures flourished for one or two decades during which schemes for radical ecological change were discussed. Most of them called into question the principle of economic growth and the achievements of consumer society. The path of (industrial) development Western societies had followed since the early nineteenth century seemed to be a big mistake. Environmentalism and political ecology became a prominent symbol by which a wide range of radical critique of contemporary society could be articulated. Environmentalism was not only a "cultural" movement but a political one as well. Since the early 1970s, it has thus been impossible to imagine politics without environmental protest and Green parties.

As environmental protection was a governmental project as well as an oppositional issue, administrative initiative and publicly articulated requests interacted, resulting in growing political activity and the creation of new institutions, legislation, and administrations. Since the 1980s, environmental policy has influenced supranational activities, such as the European Community and the Rio Conference on sustainable development. The institutional "capacities" of environmental policy grew considerably. One of the most interesting examples demonstrating the interrelatedness of governmental action, the politics of opposition, media discourse, and scientists looking for resources is perhaps the debate on *Waldsterben,* meaning "death of the forest." First discussed in 1981, the Waldsterben complex has created a momentum of discourse, research, and institutions that still works very well, answering to the interests of politicians, journalists, foresters, forest owners, and the environmental movement, although by now most of the forestry experts agree that there never was a dangerous threat to the German forests.[29]

Indeed, since the beginning of the 1970s, the scientific findings on the state of the environment have become all but univocal. The complexity of environmental knowledge no longer allows simple assessments and solutions. However, environmental protection has been integrated, since the 1970s, into the routine of everyday politics in most Western countries and on the supranational level. The debates on climate change and global warming are a pertinent indicator for the recurrence of environmental issues on both national and international levels. It shows the existence of political, administrative, and scientific "capacities" enabling the world community to address environmental problems, at least on the level of problem assessment and agenda setting. At the same time global warming denotes the limits of environmental protection when it is confronted with competing interests.

The long-lasting results of the "ecological turn" have been the political and scientific institutionalization of environmental concerns and a lasting public awareness of pollution and environmental degradation. There is still a certain consensus that environmental protection is a moral duty with respect to future generations. However, the influence of environmental knowledge on Western mentalities is limited. The impact of environmental awareness on private consumption remains insignificant or at least confined to the consumption of organic food (as a question of health) and certain environmentally friendly products. Sociologists have frequently noticed that declarations of environmental awareness in opinion polls are not linked to changing practices and consumption patterns.[30]

Environmental Protection Practice

Taking the perspective of environmental protection practice, Western societies took an important step around 1970. Since then, they have created multiple programs to lower harmful industrial emissions, be it incentives or emission limits. Nevertheless, they did not change their path of economic development.

The environmental movement had advocated a fundamental change, hoping to create a new "ecological" economy. The green schemes advocated a self-sufficient, semihandicraft, semiagrarian, almost nonconsuming, decentralized, "soft," healthy, and egalitarian society. They combined elements of agrarian romanticism, socialist ideals, and the paradigm of the "limits to growth."[31] In fact, Western societies conserved their principles of development and continued to rely on traditional pollution control strategies.

Two of these time-honored strategies in fighting pollution that continued to prevail were the so-called technical solutions and end-of-pipe technologies, which have existed since the beginning of industrialization. Technicians tried to develop existing modes of production by the use of filters or by improving the efficiency of the use of natural resources. These solutions often proved successful, as was shown by the historian Mathias Heymann, not least because they met support from various social groups, including technical experts, industrial leaders, and politicians.[32] Nearly all the successes in pollution control mentioned earlier were realized by "technical" strategies.

However, seemingly "soft" strategies have also been adopted in some areas. One of the most prominent examples is river management. After the Oder flood of 1997, West German chancellor Helmut Kohl advocated that one should give back "space" to the rivers, because humans should live in peace with nature.[33] In this spirit, recent water management schemes claim to "listen to nature" instead of elevating dikes. Newly created polders are meant to receive additional water in times of high water to prevent flood disasters. Indeed, the principle of confining and accelerating the water flow has been replaced by the principle of slowing it down. This seems to be "softer" or more "natural," but in fact the new water management is dominating nature in as hard a way as ever since industrial times. Both polders and dikes are artifacts. The creation of both damages existing ecosystems. The need to confine and to control rivers will remain crucial as long as their banks are settled areas. The human control on nature has become smarter, but it has by no means decreased its impact.

The creation of "second nature" has not diminished since the 1970s; from this perspective no redirection of the human-nature relationship has taken place. However, since the ecological turn, Western societies have developed many institutionalized instruments and strategies to assess the possible ecological costs of new developments and techniques—for example, by environmental-impact analysis. This seems to me one of the most important innovations of modern environmental policy. For instance, the construction of every new road will be evaluated by ecological expertise. Their expertise will not prevent the road nor the increase in traffic, but it can influence the road's shape. Perhaps it can prevent the destruction of a biologically important habitat or encourage the construction of a noise barrier in the vicinity of housing areas. Environmental-impact analysis does not change the general direction of technical development, but it helps to assess and to confine its ecological costs.

In some cases modern environmental policy is trying to prevent completely harmful developments; one of the most important examples in the past decade is perhaps genetically manipulated food. Whereas some nations tried to confine the problem by prohibitive legislation, it is in fact difficult, if not impossible, to prevent the advance of genetic engineering in farming, as long as its products are cheaper than traditional modes of agrarian production. Market relations remain of primary importance for the success of new technologies. There have been several innovations in environmental protection strategies since 1970, but none of them have altered the general direction of economic development, growing resource use, and increasing manipulation of the environment. Moreover, most of the instruments were well known before 1970. However, their use and enforcement have been intensified. The ecological costs of technical and economic development have been lowered in many respects by an increase in monitoring. Nevertheless, the far-reaching idealistic schemes of ecological renewal, put forward in the 1970s, have not been implemented.

THE ECOLOGICAL TURN has not caused a redirection in the "material" relationship between humankind and its environment. The ecological debates of the 1970s added a new field to society's preoccupations. However, they did not create fresh paradigms and patterns of behavior, and if so, only to a very restricted extent. The logic of economic growth, resource use, and exploitation of the environment, existing since the early industrial age, are still very much in force. Concerning the state of the environment, there are many improvements related to pollution problems or species conservation in Western societies, but there are also some alarming developments not the least on the global level. These findings neither support the idea of a state of peace between humankind and nature since 1970, nor the vision of a highly dangerous risk society. Instead, modern societies have incorporated to a certain extent institutions and instruments dedicated to mitigate the ecological costs of new technologies. But as many of them were known before 1970, and as their role correspond to other specialized institutions in modern societies, there is no need to classify them as a part of "second" or "reflexive" modernity—not least because modernity has always been reflexive.[34] Environmental history thus does not necessarily support the theory of the end of "classical modernity" by or around 1970; rather, it calls into question the soundness of this kind of short-term periodization.

Although the basic relationship between society and nature has not changed fundamentally since the 1970s, the massive politicization of environmentalism and a growing public awareness of environmental degradation are new phenomena. From the point of view of conservation history and pollution control history, there was indeed an ecological turn, characterized above all by the creation of a comprehensive idea of environmental problems and their interrelatedness in a global perspective. For a decade or so environmental protection even played a key role in public discourse. The ecological turn will therefore remain an important landmark for the periodization of contemporary environmental history. But we should bear in mind the limitations of its impact. Modern environmentalism never was influential enough to serve as a basis for an "ecological era" in the history of modern societies, despite the hopes of many partisans of political ecology.

9

The Knowledge Society

FRANK UEKOETTER

◈ "Knowledge is power." Few quotations have attracted more attention, and stirred more controversy, than the famous dictum attributed to Francis Bacon. As so often, the quotation is not an original one, as the *Novum Organum*, the treatise that provides the clearest expression of Bacon's revolutionary philosophy, does not contain a corresponding sentence. However, the general thrust of Bacon's remarks was powerful enough: his philosophy marked a departure from traditional Renaissance notions of knowledge, depicting the quest for knowledge as a path toward the betterment of society. Bacon's philosophy was essentially a blueprint for scientific research: with observation and experimentation, mandatory publication of results, and the successive formulation of general principles based on the findings, he defined key elements of proper research that hold sway to the present day.[1] With that, Bacon attracted a wide and disparate set of admirers from Comenius to Kant, a remarkable accomplishment for a person who did not develop a new theory of knowledge or make a significant discovery or invention (although he famously died from hypothermia while conducting an experiment). Controversies continue to abound on the precise nature of early modern science, but at the same time there is little disagreement that early modern Europe was the birthplace of the modern knowledge society.[2]

Is there a peculiar environmental perspective on the rise of Baconian science? In her book *The Death of Nature*, the environmental historian Carolyn Merchant has answered that question in the affirmative. "The world we have lost was organic," Merchant asserted, suggesting a shift from a holistic, female concept of nature to an exploitative, masculine view.³ As one of the earliest environmental history monographs, *The Death of Nature* received wide attention when it was published in 1980; but in retrospect the book's deficits are plain. It not only stands out for significant errors, it also presents a rather simplistic view of the environmental impact of the modern knowledge society. It is by no means certain that increased knowledge about nature necessarily leads to exploitation. Scientific research can also foster respect, and even reverence for nature. The German scientist Alexander von Humboldt may serve as a prime example.⁴ In fact, Bacon himself is far more ambivalent in that respect than Merchant acknowledges. In the third aphorism of his *Novum Organum*, the emphasis is not on exploitation but on subordination to nature: "natura enim non nisi parendo vincitur," Bacon declared ("nature to be commanded must be obeyed").⁵

Since then, countless experts have had similar experiences and expressed them in similar terms. In a 1944 speech on *The Mission of Soil Conservation*, Hugh Bennett, the chief of the U.S. Soil Conservation Service and the charismatic leader of the American fight against soil erosion, declared with uncommon modesty: "I have learned that man must work with Nature, and not against her, or sooner or later suffer the consequences."⁶ When the historian of technology Hans-Luidger Dienel studied the perception of nature by German engineers around 1900, he was surprised to find a great amount of sensitivity, and little in the way of a quest for "power over nature."⁷

Quite apparently, Merchant's reading ignores the highly ambivalent role that scientific knowledge presents from an environmental perspective. It allows a much more intensive use of natural resources, but it also allows a more comprehensive view on the environmental implications. As a result, the precise environmental impact depends strongly on the specifics of the individual case, and the parameters that one employs for the definition of an "environmental" perspective. A few examples demonstrate this point.

- In some cases the impact of scientific knowledge was clearly negative. Nuclear physics permitted the construction of bombs with heretofore unknown explosive powers. Nuclear tests during the late 1940s and 1950s led to a worldwide exposure to nuclear substances that can still be measured today. Nuclear fallout was one of the first truly global environmental problems, and concern over radiation became one of the key issues of the early environmental movement.⁸ At the same time, the much-touted "peaceful use" of nuclear power did not fare much better among environmentalists, and concerns over nuclear power plants have been a catalyst of environmentalism in many countries.[9] Advocates of nuclear energy have recently come to use the greenhouse effect to claim environmental credentials, but that is clearly a strategy to deflect attention from the hazards of nuclear energy. One of the most dangerous substances on earth, plutonium, is a direct outgrowth of this scientific research.

- In some cases the impact was clearly positive. Thanks to engineering knowledge, efficiency in the use of fossil fuels has improved dramatically over the past two centuries. The Newcomen steam engines, built in the eighteenth century to drain coal mines, were so wasteful that cynics asserted they would need a coal mine of their own for fueling.[10] As late as 1897, the Prussian Technology Council (Technische Deputation für Gewerbe) declared that the Diesel engine's superior fuel efficiency would not make it a significant competitor for the steam engine, "for there are many cases where the steam engine, while no necessity, offers so many advantages that the economic side will be neglected."[11] However, coal shortages and growing fuel prices soon induced a change of heart, and fuel efficiency emerged as a distinct field of German engineering in the 1920s. Similar trends exist in other fields of resource use. For example, the Eiffel tower in Paris, built from 1887 to 1889 with seven thousand tons of steel, could have been built with 20 percent less weight even at the time if the designers had used a more modern type of steel. Today, two thousand tons would be sufficient.[12]

- In some cases quotations from experts were open to hugely divergent interpretations. For example, there are several ways to interpret Jean Brunhes's statement of 1909 that "man has reached the limits of its cage."[13] One reading would stress that Brunhes emphasized the end of the age of exploration and the closing of frontiers all over the globe, thus asking humanity to abandon all hope for infinite frontiers. A different interpretation could point out that, given the limits of the globe, there was a need for a more intensive exploration and exploitation of resources. A third view could read the statement as more of a philosophical remark with a whiff of fin de siècle. Needless to say, Brunhes was not the only person who produced ambiguous quotations.

- In some cases expert knowledge allowed the mitigation of environmental problems that were themselves the result of scientific knowledge. For example, new notions of hygiene led to the development of elaborate water supply systems, and discussions over what to do with the resulting sewage evolved into complex expert discussions. Several issues were at stake: Should human feces be stored in cesspools, or rather be sent into a subterranean sewer network? Should there be a combined sewer, or rather separate conduits for wastewater and rainwater? Should sewage be cleaned before being released into running water, and if so, how? All these questions were subject to prolonged discussions, and although it would be incorrect to suggest a straight line of progress, there can generally be no doubt that expert knowledge has greatly reduced a lot of vexing water pollution problems over the course of decades.[14] Similarly, discussions of air pollution problems inspired many experts to develop a wide array of cleansing technologies. To be sure, the motivation was not necessarily of an environmental kind. The driving force behind the development of electrostatic precipitators, invented simultaneously by Frederick Cottrell of California and Erwin Möller of Brackwede, an industrial suburb of Bielefeld, was not only pollution control but also the collection of precious metals.[15] Some environmentalists have criticized filters and scrubbers as simple "end-of-the-pipe" technologies that are inferior to comprehensive solutions, but it is equally clear that these technologies have greatly reduced the contamination of ambient air. Today, pollution control engineering is not only a vast scientific community—the American Air and Waste Management Association has more than nine thousand members in sixty-five countries—but also a branch of industry that environmentalists frequently mention to dispel notions that environmentalism costs jobs.

- In some cases the impact of scientific knowledge is highly contested and undecided because of a dearth of experience. The case of genetically modified organisms provides a good case in point. While environmentalists are almost unanimous in their condemnation of genetic modification, the producers of genetically modified seed are claiming environmental credentials. The use of new plant varieties, they argue, will boost agricultural production, thus leaving more room for nature, and make plants resistant to certain kinds of pests, which in turn will allow a reduction in pesticides, an issue that has been on the environmental agenda since Rachel Carson's *Silent Spring*. Sentiment may induce many environmental historians to side with the former group, but empirical evidence is still too scarce to make an informed judgment.[16]

- In some cases scientific knowledge took the place of other forms of knowledge with different environmental implications. A prime example is the field of soil science and plant nutrition, where traditional knowledge about soil fertility was pushed aside over the course of the twentieth century by an ever-increasing emphasis on chemistry. Agricultural authorities were still keen to stress the importance of organic matter for plant nutrition in the early 1900s, but emphasis gradually shifted toward nutrients like nitrogen, phosphorus, potash, and lime. Over time the soil came to be seen as merely a temporary storage for nutrients on their way from the chemical factory to plant matter.[17] This growing emphasis on nutrients induced a massive increase in mineral fertilizer use, resulting in groundwater contamination and eutrophication of the landscape. But does that make traditional agriculture more environmentally benign? The cultivation of wasteland was a prominent part of eighteenth- and nineteenth-century agriculture because of the low productivity of traditional farming and the constant fear of starvation. These wastelands were (and are) often rich with biodiversity and rare species. With the use of scientific knowledge, farm productivity increased to such an extent that famines became a distant memory in Western societies, making the quest for additional farmland pointless. Thus prescientific agriculture was not simply "better" from an environmental standpoint but merely "different."

- In some cases the environmental perspectives are so conflicting as to cast doubts on the wisdom of the entire question. The chemical industry provides a case in point. On the one hand, chemical plants have a well-documented record as polluters of air, water, and land.[18] During the environmental era numerous incidents from Love Canal to Bophal served to demonstrate the hazards of industrial chemistry. On the other hand, the chemical industry has enormous potential for reusing waste materials, thus making it a pioneer of recycling long before the word came into use.[19] Some products of industrial chemistry have been shown to be cancerous, or otherwise harmful, while others have had a positive impact in that they have helped to conserve resources or replace a more dangerous product. In fact, some products have qualified in both respects. For example, when tetraethyl lead was introduced as a fuel additive in the 1920s, manufacturers touted it as a welcome means to boost fuel efficiency in the wake of widespread fears of an exhaustion of petroleum reserves; decades later, however, environmentalists chastised the same substance for its toxicity.[20] Clearly, any evaluation of the environmental impact of the chemical knowledge society depends strongly on which substances and which aspects one chooses to stresses.

With these examples one thing should be obvious: the environmental impact of the modern knowledge society is a complicated one. The picture looks different from case to case, depending strongly on the specifics of the individual topic and the criteria that one employs. In a way one may even call that a welcome finding, as it opens new fields of inquiry for environmental historians, and it should not go unmentioned that there are still numerous topics waiting to be explored in the no-man's-land between environmental history and the history of science. Of course, one may entertain some hope that these future studies might clarify the situation a bit more and provide some firmer basis for a general assessment. But for the time being, it seems that it may be useful, in the quest for comprehensive turning points in the environmental history of the modern knowledge society, to focus not on the *content* of knowledge and its environmental impact but on the *structure* of knowledge. Although the inclusion of scientific experts has played out in widely different ways, the fact that experts became involved marked a turning point in its own right.

A few examples may serve to demonstrate this point. The American debate over the urban coal smoke nuisance started off as a civic crusade, with upper-class reformers, usually organized in civic associations or even special smoke abatement leagues, urging city councils and mayors to pass and enforce smoke ordinances. But the movement's character changed significantly when city governments started to employ mechanical engineers as municipal smoke inspectors. On the surface the main effect was a change of strategy: instead of vigorous prosecution, formerly the common (and mostly ineffectual) method of choice, smoke inspectors favored a cooperative approach in that they attacked smoke as a symbol of waste, offering help as firing consultants to smoking businesses and generally seeking to enlist the support of the business community in the fight against smoke. But the real impact went deeper than that. From now on, everyone concerned about smoke had to compete for attention and authority with a self-confident group of experts, organized in a nationwide Smoke Prevention Association since 1907.

To be sure, the civic reformers were not generally hostile to the growing importance of engineers. On the contrary, they were usually the driving force behind the employment of engineering experts, and some civic leagues even hired engineers on their own expenses. But the general structure of the drive against smoke had changed permanently. Citizens now had to cooperate with smoke inspectors and, with the coal smoke discourse being dominated by engineering issues to an ever-increasing extent, often found themselves taking a backseat in the fight against smoke. In short, it turned out that there was no way back to a purely civic drive against smoke. Citizens could refute the experts' claim to authority, but they could no longer do without scientific expertise.[21]

A similar development took place in the field of agriculture. In Germany scientific studies of agricultural issues go back to the early nineteenth century, and some countries, most notably England, can claim an even longer tradition of research.[22] But for a long period of time scientific experts assumed a rather modest, subsidiary role, seeing themselves as occasional advisers to practitioners in the field, rather than as masters of the agricultural craft. As late as 1906 the head of the Darmstadt agricultural experiment station, Paul Wagner, one of the leading authorities on fertilizer issues of the time, refused to provide farmers with specific instructions on the use of fertilizers. "You will not bother me with such a silly request," he declared. Instead, he asked every farmer to check the fertilizer needs of his fields himself.[23]

A century later the situation has changed dramatically. Chemical analyses of soil samples, performed by chemists at agricultural experiment stations, are the dominant mode of perception when it comes to the fertilizer needs of a certain plot of land.[24] In fact, the agricultural science establishment has made itself indispensable to such an extent that even reformers find it inevitable to draw on their expertise. When Renate Künast became the German secretary of agriculture in the wake of the mad cow disease scandal in early 2001 and pledged to reform German agriculture from head to toe, the ministry's scientific council resigned en bloc, only to be reconvened a few weeks later with the same chairperson and most of its former members.[25]

Even nature conservation has evolved into a profession of its own, with biologists usually providing the lion's share of expertise. Biologists have played a role in the nature conservation movement from its inception, but the movement's constituency was much larger for a long period of time. Great Britain's Royal Society for the Protection of Birds was led for more than half a century by a noblewoman, Winifred Duchess of Portland, while Ernst Rudorff, the spiritual founder of the German Heimatschutz movement, was a professor for piano at the Royal Institute of Music in Berlin. Avraham Yoffee, the first chair of Israel's Nature Reserves Authority Council, was a general.[26] Even those who had made a lasting contribution to nature writing were not necessarily eminent researchers. For example, Henry David Thoreau was "not a great naturalist," as the biologist

Edward O. Wilson noted, with apologies, in a recent book.[27] But over time the situation changed, a process that is one of the most understudied developments in the history of the environmental movement. Presumably, institutionalization and bureaucratization have played a prominent role. Few civic leagues bother to check the university diplomas of their members, but the situation is different with public agencies, and academic degrees are often mandatory requirements for administrative jobs.

As a result, it was often officials who were complaining of unprofessional conduct in conservation affairs. As early as 1934, Walther Schoenichen, the head of the Prussian Agency for the Protection of Natural Monuments, published an article that criticized citizens who came into conservation with "a certain superficial, amateurish and unprofessional idea of the concept of nature protection."[28] A second important reason was probably the growing importance of species protection during the second half of the twentieth century, for identifying species and their living conditions is a prime qualification of trained biologists. The crisis of confidence that nature conservation is facing in many countries may well be related to the fact that conservation experts often speak in a language that laypeople do not understand.[29]

However, it would be shortsighted to depict expertise as merely a logical outgrowth of scientific work. The historian of science Peter Lundgreen has argued that the rise of the modern professions, the rise of science, and the rise of academic institutions were originally three separate trends that gathered momentum only over time as they began to interact and mutually reinforce each other to an ever-growing extent.[30] Following Lundgreen, one may argue that a turning point in the history of the modern knowledge society is in fact a cluster of three separate, although closely interrelated, trends: (1) the growing importance of academic expertise; (2) the growing strength of professions and professional associations; and (3) the creation of formal institutions for research and education. But that makes the quest for clear periodizations even more complicated. When did these three trends gain momentum?

Few countries make it as easy as the United States, where the experts' rise to power can be assigned quite clearly to the fifty years from the late nineteenth century to the 1930s. The starting point was the Progressive Era, which saw "America's first full-scale attempt to come to terms with the rapidly emerging multicultural, urban, industrial society."[31] In his influential article on the Progressive Era, the historian Daniel Rodgers has argued that the political philosophy of progressivism was characterized prominently by a language of "efficiency, rationalization, and social engineering."[32] Efficiency became a universal buzzword—in a famous phrase, Samuel Hays spoke of a "gospel of efficiency"—and in many cases the implication was that only the knowledge of professional experts could assure an "efficient" approach.[33]

It is indeed remarkable how the federal government came to provide experts with an unprecedented new status within a matter of years. Although there had been a Division of Forestry within the federal government since 1879 (a date that coincides nicely with the formation of the American Forestry Association only four years earlier), it did not move aggressively into the business of forest management until Gifford Pinchot became chief of the division in 1898. In 1907 the Reclamation Service became independent from the Geological Survey and set out to manage the scarce water of the American West in a more efficient, "multipurpose" way. On the local level sanitary engineering emerged as a distinct profession, with such charismatic figures as George E. Waring Jr., New York City's commissioner of street cleaning, in the vanguard.[34] The trend toward institutionalized expertise languished during the laissez-faire 1920s, only to gain a second push during the New Deal. When the Dust Bowl struck the semiarid regions of the Midwest, the Soil Conservation Service grew into a huge bureaucracy with more than ten thousand members within a matter of years.[35]

However, the rise of professional experts was a more prolonged process in other countries. For example, the creation of the U.S. Forest Service occurred at a time when professional foresters were already a fixture of German policy for decades. Around 1770 several German states had taken initiatives to create Forest Academies (*Forstakademien*) or other institutions for the systematic training of foresters, which gradually transformed foresters from a much neglected and despised group into a respected profession.[36] Germany's academic forestry was a pioneering effort on a global scale, and German-trained foresters pursued careers in a wide range of countries from India to the United States. However, it was not an effort that defined a path that other professions would follow quickly. After all, foresters profited from two peculiar conditions. First, their profession met nicely with the interests of the late absolutist state in the proper management of a state-owned resource: with well-trained foresters, the states hoped to increase per-acreage yields, thus allowing a greater sale of wood or the safe supply for profitable but notoriously fuel-thirsty enterprises like salt- or metalworks. Second, the rise of academic forestry was ultimately a product of the "wooden age," where forest products were still a crucial resource whose abundance (or lack thereof) decided society's well-being. Thus forestry gradually lost its prominence when society's thirst for energy shifted from wood to coal and oil in the wake of industrialization.

But some professions rose significantly later in Germany than in the United States. While America had an association of air pollution control engineers since the Progressive Era, German engineers waited with the formation of a distinct clean air committee (Fachgruppe Staubtechnik) within the powerful Society of German Engineers (Verein Deutscher Ingenieure) until the mid-1920s, and this group remained little more than a noncommittal meeting place for decades. It was not until the mid-1950s that the Society of German Engineers took the initiative to set up an advisory body for government consultation, a role that America's Smoke Prevention Association had assumed almost from its inception.[37] Curiously, it had been the strength of the German bureaucracy that had forestalled an earlier formation of these bodies similar to the United States. In Germany the mechanical engineering profession emerged in constant conflict with the bureaucracy because of the latter's insistence on judicial expertise for all decision-making posts. As a result, mechanical engineers saw strict air pollution laws primarily as an undue empowerment of technologically incompetent jurists and generally sought to keep the government out of their business as much as possible.

It is important to recall these divergent chronologies, for they caution against simple "challenge and response" interpretations in the environmental history of the knowledge society. Industrialization may have made society more complex and, with the use of new technology, may have increased the need for expert knowledge, and yet it is by no means certain that expertise was actually used and institutionalized. Even when professions did exist whose knowledge had a potential for mitigating certain problems, it was a long and complicated process to bring this knowledge to bear, and this process was inevitably beset with the vagaries and uncertainties that any prolonged political process implies. It is only from a broader perspective that the huge momentum of the scientific revolution becomes clear. In the long run the trend toward institutionalized research was irresistible—and it is to the present day. In fact, states all over the world found expertise such a precious resource that they sought to monopolize it as much as possible.

The environmental history of the twentieth century is full of big bureaucracies that control their field through the centralization of knowledge; mining and hydrology may serve as examples here. In a way it was a bizarre reinterpretation of "knowledge is power"—a shift of focus from the first to the second noun. Where Bacon stressed the need to generate new insights, states found that when it came to the tasks of the modern age, power presupposed control of knowledge. In light of the huge size of expert bureaucracies, it is easy to forget that these empires could not come into being on their own. It usually took a strong social consensus to build these expert systems, and institutionalizations were typically the strongest when problems were touching strong norms and values.

Cleanliness was a quintessential bourgeois value, and that may explain why urban sanitation became such a prominent issue in all Western countries when the growth of urban agglomerations made sanitary conditions precarious in the late nineteenth century. Major epidemics underscored the need for change. The last cholera epidemic on German soil claimed some eighty-five hundred lives when it struck Hamburg in 1892.[38] As a result, professions developed much quicker in the field of urban sanitation than in the fields of air pollution or noise control. The German Public Health Society (Deutscher Verein für öffentliche Gesundheitsplege) was organized in 1873 as a platform for discussions on all aspects of urban sanitation. When the society took a stand in favor of combined sewers, a dissident association evolved, the International Association for the Sanitation of Water, Land, and Water (Internationaler Verein gegen Verunreinigung der Flüsse, des Bodens und der Luft) in 1877. However, the name was somewhat misleading: the International Association was really a German association that actually focused much more narrowly on water issues than the German Public Health Society. But while the latter society thrived, the International Association never managed to move beyond a marginal role. By 1900 the association had essentially become a one-man-crusade of Georg Bonne, a physician from Hamburg.[39]

The case of the International Association demonstrates that a discussion of turning points should not remain confined to the period of foundation of the modern professions. In fact, professions have demonstrated an enormous potential for change over time. When engineers moved aggressively into smoke abatement during the Progressive Era, citizens saw them as a welcome, and in fact dearly needed, support in the fight against an urban plague. But half a century later, the same group of experts was merely a rubber stamp for voluntary efforts of industry against smoke and other pollutants, isolated from the growing environmental sentiment in postwar America. When the Reclamation Service was set up in 1907, it was clearly an innovation over the riparian use of water in the West, and Progressives saw it as a guardian of the common good. A generation later the same institution was a closed-off group of experts who pursued dam building as a mostly self-serving enterprise.[40] The general pattern was basically the same. As the professions gathered momentum, they gradually became estranged from general society, instead seeking to appease a more limited constituency: industry in the case of smoke inspection, and politicians and irrigation farmers in the case of the Reclamation Service. Confident that they knew how to do "the right thing," the professions evolved into a distinct political sphere where the common rules of democratic society no longer applied.

It seems that the modern knowledge society reached the apogee of its prestige and power around the mid-twentieth century. To be sure, it was not an unambiguous climax, for the horrors of World War II were still vivid memory. Shocked by the corruption of science and engineering under National Socialism, German engineers philosophized about a "curse of technology" in the 1950s. In the United States, Hiroshima and Nagasaki, along with the ensuing conflicts over governmental control of nuclear research, symbolized the deep moral abyss of modern scientific knowledge. And yet nuclear physics had come to nourish utopias in a way that few other technologies ever did: a coming "atomic age" would bring prosperity to all people and an end to vexing energy troubles. The widespread euphoria becomes even more remarkable when one notices that, unlike the "electric dreams" around 1900 or the buzz about computers and the Internet in recent decades, it was not rooted in an existing technology. The "atomic age" reached its peak at a time when there was not a single commercially operating nuclear power plant in the world.

As a result, professional pride was reaching unprecedented levels, culminating in fantasies of a caste of nuclear scientists as the high priests of the atomic age. But two decades later the nuclear community would symbolize something altogether different: the arrogance of the experts who had become neglectful of the inherent risks of nuclear power in the pursuit of their goals.[41]

Against the background of the history of the modern knowledge society, one cannot help but call the rise of environmentalism a highly unlikely event. With the system of professions firmly in place, and with the benefits of their expert knowledge, from growing life expectancies to unprecedented affluence and mobility, apparent for everyone in the Western world, the modern knowledge society might have appeared as an invincible force. But at its core every expert system was based on the trust of society, and this meant that even huge scientific empires were inevitably resting on a shifting foundation. After all, trust is a precarious resource in modern society, and trust can decline, or even be called off completely, at any time.[42]

That is what happened in the late 1960s and early 1970s. Many citizens chose to no longer trust the ruling professions and cancelled the informal contract that was their power base. In some cases trust was lost within a matter of days. In the conflict over the nuclear power plant project in Wyhl in southwest Germany, for example, a key turning point was when the utility invited an expert to a hearing for villagers, many of them winegrowers, who declared that the clouds from the nuclear facility's cooling towers would not affect their business because vines did not need too much sun. For a community that took pride in the warm, sunny climate that made their wine so delicious, that was a turning point: they now knew what to think of "expert talk."[43] The modern knowledge society thus stood on shaky ground in social respect. Obviously, it was much more difficult to gain trust than to waste it. But the uncertainties were also of an empirical nature. Key arguments of the antinuclear protests could be gleaned directly from expert reports. Critics have argued that nuclear power was the product of an iron triangle that shielded it from opposition; but that notion has been disproved as too simplistic.[44]

Some environmental historians have been even more skeptical, arguing that experts were simply advocates of industry who helped to shield the vested interests from environmental qualms. But such a reading clearly ignores the dual loyalty of professionals, who subscribe not only to the interests of their employers but also to professional standards of "good practice." But helpful as these expressions of professional honesty were for critics of nuclear energy and other environmentalists, they were no substitute for the environmentalists to develop their own expertise. As the sociologist Peter Weingart has noted on the modern knowledge society: "There is no way to replace science through another system. There is no other source of secure, certified knowledge."[45] So environmentalists could not do without experts altogether, and it is impressive indeed how the antinuclear movement reached a level of technological sophistication that was a match for the nuclear-industrial complex. It certainly helped that experts like Klaus Traube switched sides during the dispute, but books like Holger Strohm's *Friedlich in die Katastrophe* clearly showed an impressive ability to master even obscure technological details.[46]

For a time the experts grouped neatly into camps, and the nuclear controversy was by no means unique in that respect. Shortly after Egmont Koch and Fritz Vahrenholt wrote *Seveso ist überall* (Seveso Is Everywhere)—an indictment of the hazards of the chemical industry in the wake of a serious industrial accident in Seveso, Italy, in 1976—the trade association of the chemical industry published a counterpublication, creatively entitled *Seveso ist nicht überall* (Seveso Is Not Everywhere).[47] But the two camps soon disintegrated into a wide spectrum of experts, and the environmental knowledge society has attained a level of complexity nowadays that makes it difficult for many environmentalists to develop an informed opinion. "There are very few black hats or white hats out there," one author remarked on the experts dealing with toxic and hazardous waste in the

1990s. The general impression, surveying specialists from diverse background, was "a sea of gray hats."[48]

With that it is hard to avoid the conclusion that the environmental movement is standing on shaky ground. Gone are the days, at least in much of Western society, where one could simply rely on one's senses to define priorities. For example, when Great Britain's Parliament discussed water pollution on the Thames River in the mid-nineteenth century, there was no need for expert witnesses to learn about the problem's character or extent because the problem, the famous "Great Stink" of 1858, lay literally at its doorsteps.[49] Nowadays, it is nearly impossible to evaluate environmental problems without the consultation of experts because problems like radiation or chemical pollution or climate change defy a simple intuitive approach. So how does one define priorities on environmental issues?

When Bjørn Lomborg—a professor at the University of Arhus in Denmark who gained worldwide fame with his controversial book *The Skeptical Environmentalist* of 2001—invited a number of experts to a four-day meeting to choose those environmental problems that deserve the greatest attention, the result was inevitably controversial. Staunch criticism came from activists in the fight against the greenhouse effect, a problem that Lomborg's group had placed at the bottom of its list.[50] But this criticism may have had more credibility if it were not well known that climate change has evolved into a multimillion-dollar business of its own, with researchers, administrators, and many activists having a personal stake in keeping the climate debate alive. The modern knowledge society has finally become aware of its own ambiguities, and once again there is no way back. Today's activists know that the search for a truly autonomous, objective expert remains elusive, as it has always been.

Who knows, maybe future environmental historians will see the "great environmental awakening" as merely a "great deception" because it focused on the wrong issues? It would not be the first ambiguous turning point in the environmental history of the knowledge society.

10 Desertification

ALON TAL

Desertification is among the most misunderstood—and the most neglected—of the world's global environmental challenges. Images of irrepressible waves of sands overwhelming civilization are not entirely fictitious. Indeed, the spectacle can be witnessed every day from the Sahara to China; frequently, natural phenomena can lead to desertification.[1] For the most part, however, desertification refers to the much less dramatic but far more pernicious steady decrease in land productivity that takes place in drylands. It is important to emphasize another misconception. Although they may contain productive oases or river valleys, true deserts (arid and hyperarid lands) are typically *not* the areas in which desertification on a large scale takes place. Rather, it is in the semiarid and subhumid drylands that receive low, often seasonal rainfall and support soils with modest organic content, where the relatively low-land productivity may decline even further.

When the nations of the world finally negotiated a treaty to "combat" desertification in 1994, they defined the phenomenon as "land degradation in arid, semi-arid and dry sub-humid areas resulting from various factors, including climatic variations and human activities."[2] A higher resolution definition put forward by the United Nations Environmental Program (UNEP) characterizes desertification by five processes that damage land productivity: (1) vegetation degradation, (2) water erosion of soils, (3) wind erosion of soils, (4) soil salinization, and (5) soil compaction—in the drylands. Natural shifts in climate or meteorological processes can surely contribute to these processes. For example, geologists believe that the change in wind directions after the Holocene period replaced the plentiful deposition of Sahara loess in Israel's Negev Desert, with smaller quantities that arrived from Saudi Arabia. With replenishment diminished, loss of soil due to natural erosive processes was inevitable.[3] But generally desertification refers to the loss of soil productivity that is driven by such anthropogenic activities as deforestation, overgrazing, or poorly considered water management.

There is nothing new about these activities. Farmers have been aware of the vulnerability of their lands to human abuse and of phenomena like irrigation, waterlogging, soil salinization, or riff and gully erosion from time immemorial. And the ancients were not without effective responses. The Old Testament is full of stories that refer to the importance of imposing stock limits in rangelands (as in the pasture distribution between Abraham and Lot) or normative prescriptions, such as the requirement of crop rotation and sabbatical years for soil rejuvenation.[4] Their terraces still define the gnarled landscape of Israel. Although it is not clear that the farmers of old were fully cognizant of desertification processes when adopting such practices, and their implementation alone cannot guarantee the prevention of desertification, they can meaningfully contribute to sustainable land management. Jared Diamond, a Pulitzer Prize–winning ecologist, described the process that undermined many a civilization in the ancient Near East: "Because of low rainfall and hence low primary productivity, (proportional to rainfall), regrowth of vegetation could not keep pace with its destruction, especially in the presence of overgrazing by abundant goats. With the tree and grass cover removed, erosion proceeded and valleys silted up, while irrigation agriculture in the low-rainfall environment led to salt accumulation. . . . Thus, Fertile Crescent and eastern Mediterranean societies had the misfortune to arise in an ecologically fragile environment. They committed ecological suicide by destroying their own resource base."[5]

But "desertification" only emerged as a salient term of reference and a modern international problem during the second half of the twentieth century. As early as 1927, a French biologist working in Tunisia documented the low productivity of rangelands there.[6] Twenty-two years later, the French colonial forester A. Aubreville coined the term "desertification" itself (in French) when he described the grim situation in West Africa: "The closed forests are shrinking and disappearing, like evaporating spots. The trees of the open forests and savannas become more and more spaced out. On all sides, the bare skin of Africa appears as its thin green veil of savanna burns releasing a grey fog of dust into the atmosphere. Arable land is carried away by the yellow waters of flood. Slabs of sterile truncated soil, bearing tufts of grass around uprooted bushes, recall a kind of leprosy that is spreading over the face of Africa."[7]

It would take another twenty years, and considerable human misery, before the world woke up to the extent of the challenge, and it was enormous. The 2005 Millennium Ecosystem Assessment reported that the "number of people affected by desertification is likely larger than any other contemporary environmental problem."[8] Desertification is therefore a relatively modern environmental phenomenon even though its symptoms have been well documented in the past—from ancient Greece to Ottoman Palestine. Part of the reason for the present acceleration of desertification processes is the link between environmental damage and the exceedance of human carrying capacity on the relatively fragile drylands. Only recently have the sheer numbers of human beings on the planet created extreme population pressures on a global scale. Several salient events and activities over the past century have addressed this global challenge. These can be divided into two categories: (1) natural disasters that have raised public awareness about the severity of the problem; and (2) more recently, international or local efforts that have demonstrated that this trend need not be destiny; just as humans have created the desertification crisis, humans can solve it.

Unfortunately, there are far more examples from the natural disaster category than the latter. The international community that has mobilized to address several insidious and vexing global environmental problems (such as the depletion of the ozone layer, whale extinctions, or trade in endangered species) has found neither the resources nor the political will to make a serious commitment to changing the planet's desertification profile. Unlike most environmental challenges that require proscriptions on development and cooperative restraint in human interaction with the global commons, the remedies required to stop and reverse desertification are quite the opposite in nature. They require thoughtful investment in sustainable development in the drylands of developing nations. This by its very nature is a gradual process, not given to dramatic turns. At the same time, unless the resources and stamina for such a sustained process are garnered, desertification will continue to be an environmental orphan that spirals out of control, leaving agricultural failure, famine, and refugees in its wake.

Land Degradation and the Dust Bowl of the 1930s

Although desertification is a phenomenon largely associated with Africa, perhaps the first signs during the twentieth century of the potential scope of devastation produced by poor soil stewardship occurred in the United States. The Dust Bowl was actually a series of dust storms that ran through the Great Plains of the United States and Canada beginning in 1933. The background to this disaster was twofold, with the first cause being anthropogenic: the overproduction associated with World War I and the accessibility of mechanized farm equipment. The second was meteorological. Farmers in the American Midwest were encouraged to develop marginal lands to meet the demand for food and fiber that the Great War and its aftermath had created. At the same time many Midwest farms became "overcapitalized" relative to the actual carrying capacity of their lands. To cover the expenses of tractors, listers, and a new generation of plows, farmers tried to press as much land as possible into production. When the United States fell into the Great Depression at the end of the 1920s, prices for food dropped and farmers had to cultivate even more marginal lands to pay their debts. Once these lands were stripped of the natural vegetation, they either fell out of production or the continuous plowing exposed the soil, making it vulnerable to the ferocious winds of the American Midwest. Damage might not have been as acute had these events not coincided with a period of prolonged drought.

The storms began in November 1933 in South Dakota, with the most famous tempest occurring on May 11, 1934. A Nebraska physician recorded: "Wind forty miles an hour and hot as hell. Two Kansas farms go by every minute."[9] The soil blew clear across the United States. In Chicago four pounds of soil fell from the sky like brown snow in summer. When the storm reached Washington, D.C., members of Congress left the U.S. Capitol building to witness the opaque clouds of dust that darkened the noon-hour skies. The snow that fell that winter in New England was red. Dust storms continued, culminating in Black Sunday on April 14, 1935—also known as the "black blizzard" of the period. The agronomic results of three years of drought and the relentless dust storms were catastrophic. Homes were destroyed as the deluge of dust covered entire towns. Crops failed in successive years and farmers could not pay back loans, leading to eventual foreclosure. Hundreds of thousands of environmental refugees left the region in despair. They became known as "Okies," after the 15 percent of Oklahoma residents who were forced to leave their lands in destitution. Eventually millions more would migrate west, having lost all prospects of prosperity on their eroded farms. (John Steinbeck's classic novel *The Grapes of Wrath* movingly describes one family's ordeal.)

Although the Dust Bowl preceded by many years international recognition of a desertification crisis, its historical significance as a turning point in the field is twofold. First, the sheer magnitude of the disaster raised public awareness about the vulnerability of soil and the potential ramifications of imprudent land management. Second, the upshot of the disaster was positive: the erosion crisis spawned institutional and legislative change in the United States. The Soil Erosion Act passed in April 1935, and President Franklin Roosevelt established the U.S. Soil Conservation Service (SCS) with a mandate to aggressively deal with land degradation.[10] This SCS remains a massive agency to this day, whose efforts have reduced desertification throughout the prairies and the drylands of America, even as dust-bowl symptoms reappeared in the United States during the 1950s and 1970s. Other nations soon followed the American lead, with soil conservation emerging as a major academic discipline. Laws and institutions have been established regarding drylands from Australia to Spain to assist local farmers to sustainably cultivate their lands.[11] When the international community began to seriously consider the issue of desertification, the know-how for addressing it largely existed.

A Population Bomb Explodes in the Drylands, 1950–1970

As civilization has advanced, the sheer number of human beings on the planet has grown exponentially. For example, 90 percent of the increase in human presence since the beginning of time has taken place during the past 350 years. But this understates the astonishing rise in population that has taken place on Earth beginning in 1950—in particular, among the dryland, developing nations. To demonstrate the significance of the phenomenon, it is instructive to divide human history into fifty-year segments. Between 1850 and 1900 the planet's population is estimated to have increased by 400 million people (from 1.2 billion to 1.6 billion people), with an annual growth rate of 0.6 percent per year.[12] During the next fifty years, which included two world wars, the population rate rose by a mere 0.2 percent (to 0.8 percent), reaching just over 2 billion people. But no sooner had the explosions from the world wars died down than the "population bomb" began to detonate. During the next twenty-five years, the rate of population growth more than doubled, averaging 1.9 percent, leaving the world with 4.1 billion people in 1975.

Developing nations, especially in Africa, were among those who saw the greatest surge in population size. Afghanistan, Burkina Faso, Burundi, Chad, China, Congo, Egypt, Ethiopia, India, Jordan, Kenya, Mali, Niger, and Nigeria—all countries dominated by their drylands—also showed population growth rates of 2 percent and higher. As villages swelled, traditional land resources became inadequate and new marginal land was put into production. It was only a matter of time before an ecological price would have to be paid. As desertification became a major scourge through African and Asian drylands, the proximate cause of any given disaster was typically drought. There is anecdotal evidence that in some cases local population growth can create a critical demographic mass for the generating of "alternative livelihoods," which may ease land pressures.[13] But on the whole the steady increase in population growth from 1950 until 1970 left little latitude for traditional adaptive responses and made the consequences far more painful.

The 1968 Famine in the Sahel

Although very much a dryland region, the Sahel also constitutes a natural greenbelt located south of the Sahara that protects pastoralists and farmers from the great desert. Although it has low natural rainfall, the Sahel has always been able to support a variety of flora and fauna along with a diverse range of human lifestyles. Deserts also have a rich variety of flora and fauna, but conditions in the Sahel have traditionally supported pastoral as well as farming communities. Elders living in the region today can remember a time when the Sahel was home to rich ecosystems that provided abundant game, including antelope, monkeys, wolves, foxes, and even elephants![14] Although rainfall was never plentiful, shepherds could find forage for their flocks and millet grew well. All this changed, however, when extensive slash-and-burn agriculture began to decimate the natural woods and bushlands.[15] It was as if population density had crossed a critical threshold and the land could no longer support the traditional farming practices. These often included an inclination to clear fields after harvest, exposing the topsoil to powerful winds. When drought set in around 1968, farmers from Mauritania in the west to Ethiopia in the east, as well as the local pastoralists, initially assumed it was just another cyclical dry year from which they would soon recover. But the usual rains did not return until 1972, and 250,000 people died because of the famine and associated disease.

When the rains did come at last, much of the soil that should have supported the recovery was gone, literally blown away. The desiccated lands, lacking the original organic matter, could not sustain the grasses and bushes that normally fed livestock. What had once been a "nondesert dryland" had become desertified. Heartbreaking images of starving children, barren and abandoned villages, and an unforgiving desert smothering once fecund lands filled television screens around the world, for the first time giving desertification an African connotation that has not changed to this day. Environmental disasters can be considered turning points if they engender a meaningful public policy response. In the case of the Sahel drought, world public opinion was mobilized and the UN General Assembly called for a global response to the crisis. The best it could do was to convene a conference that eventually was held in Kenya. The United Nations Conference on Desertification (UNCOD) brought ninety-five nations together and produced an agreement that called for nonbinding action programs. In retrospect, however, the effort engendered little action indeed.[16] But it did turn the issue of desertification into an international environmental issue, and ten years later a more effective international instrument would go into effect.

Drip Irrigation in the 1970s and 1980s

Although desertification is largely a result of human activities, human innovation also holds the potential for a solution. No invention has been more significant in revolutionizing dryland agriculture than drip irrigation. During the 1930s water engineer Simcha Blass helped to build water systems for new Jewish settlements in Palestine. A friend showed him a giant tree near his house that had flourished as a result of a leaky pipe, thriving on a steady stream of tiny droplets. Blass had an epiphany that he later compared to "a mosquito in the mind of Titus the evil." But it would take more than twenty years, during which time he ran the Water Department for the young state of Israel, before he could find time to develop an effective application.[17] By then, plastics were available and he could design the world's first drip irrigation system. Rather than flooding a plant's root zone, water is spoon fed to trees and plants, drop by drop, through narrow (traditionally) black pipes whose drippers regulate the amount of water released. Over time the systems could be linked to computer systems that optimize the rate and timing of applications.[18] In 1965, Blass sold his invention to an Israeli kibbutz that created the Netafim company and production began.[19] Forty years later irrigation was an eight-hundred-million-dollar business in Israel, and modern dryland agriculture was never the same.

The agronomic benefits are many. Fertilizers and wastewater can be delivered efficiently. Evaporation, a key factor in the creation of drylands, is greatly reduced relative to sprinkler or flood systems. Even steep terrain and shallow soils, always a problem for irrigation in the drylands, responded favorably. It was now possible to grow a range of crops in coarse sands and clays.[20] From the perspective of desertification, the chronic salinization that accompanied the evaporation of flood irrigation waters was largely avoided. And most of all, farmers were producing more "crop for the drop"—a good deal more! It was this invention, more than any other innovation, that led to Israel's impressive growth in agricultural production. While its population grew sevenfold during the first fifty years of statehood, agricultural production in a nation whose territory is 95 percent drylands increased seventeenfold. Use of freshwater in the agricultural sector, however, dropped as the new technology could effectively utilize recycled wastewater. Lands that were otherwise unprofitable, that would have fallen out of production, could now be sustainably cultivated, even in hyperarid regions. Centuries of degradation in Israel's semiarid and arid lands could now be reversed.[21] Although establishing drip irrigation systems requires substantial capital costs, its exportation to dry developing countries from Tanzania to Afghanistan has begun to change the local agricultural profiles

Not all drivers of desertification are given to a "technological fix," however. Land tenure systems, demographics, and even grazing may require policy responses of an entirely different nature. Yet there is little doubt that when science and technology are creatively used in the battle to combat desertification, substantial benefits and progress can be anticipated. Drip irrigation's steady dissemination around the world during the past thirty years demonstrated this and offered an empirical basis for optimism.

The Disappearance of the Aral Sea and Lake Chad, 1970s–1980s

In 1960 the Aral Sea was the fourth largest lake in the world, larger than the area of Belgium and the Netherlands combined. The surrounding watershed is largely defined as a drylands region. But that year it began to shrink. The sea links Kazakhstan in the north and Uzbekistan in the south, and for many years its fish provided a livelihood for the surrounding communities. During the 1950s Soviet ships could boast an annual catch of forty-eight thousand tons of sturgeon, carp, and bream.[22] It was hard to imagine that the Aral Sea was facing one of the twentieth century's greatest ecological disasters.

As early as 1918, the newly formed Soviet Union decided that it would be well to tap the two rivers that fed the Aral Sea (the Amu Darya and Syr Darya) to irrigate a variety of crops. A series of diversions were completed by 1960, by which time some fifty cubic kilometers that used to reach the Aral Sea was watering fields throughout the surrounding areas.[23] The agricultural strategy was at first ostensibly successful. Cotton production doubled between 1960 and 1980, when 85 percent of the area's farms were growing the ever-thirsty plants. But the environmental ramifications soon proved disastrous. The sea level fell by twenty centimeters a year. Then the drop began to grow geometrically, so that presently the waterline retreats almost a meter a year.

The disappearance of the sea was hardly a surprise to the Soviet government planners. Indeed, in 1968 a local engineer saw the process as inevitable. But it is a sorry sight today: what remains of the once grand water resource is ten hypersaline pools that hold a mere tenth of the original water volume. The cities built on the banks are often as much as 150 kilometers (about 93 miles) away from any water, surrounded by the dusty remnant of what was once a lovely lake. The phenomenon continues, with an area twenty times the size of Manhattan added each year to this new desert.[24] The ecological law of unintended consequences quickly began to set in. The lands surrounding the Aral Sea are now heavily polluted with a variety of metals and chemicals, hence the resulting dust storms that translated into chronic lung morbidity in the surrounding populations, producing alarmingly high cancer rates.[25]

On the African continent Lake Chad was facing a similar fate. During the first half of the twentieth century, Africa's fourth largest lake, historically shallow, was still able to provide water for twenty million people in Cameroon, Chad, Niger, and Nigeria. Never deeper than seven meters, it was always reliant on the Chari River, which provided some 90 percent of the lake's water. The river meanders for 950 kilometers (almost 600 miles) and drains a watershed of 548,747 square kilometers throughout Central Africa, before reaching Lake Chad. But while rainfall levels were dropping, more water was diverted from the river, its tributaries, and from the lake itself to support agriculture. Even faster than its Asian companion, Lake Chad began to disappear.[26] To be sure, given the flux in its watersheds, the lake had almost dried up before, in 1908 and again in 1983, before bouncing back. But the recent drop looks more ominous and irreversible. During the 1960s, Lake Chad covered more than 26,000 square kilometers, but by the 1990s it had lost 95 percent of its surface area. In a small corner of the southern basin, 550 square kilometers are still wet, but not for long. And where there had once been water and aquatic habitats, there is now desiccated and unproductive land.

Part of the problem with desertification has always been the difficulty in documentation. The steady disappearance of these two extraordinary resources—the Aral Sea and Lake Chad—was quite literally captured in aerial photographs and the attendant land degradation offered a sober wake-up call for the world. Terms like "sustainability" suddenly took on a far greater immediacy, and the desertification process that ensued had a disturbing poster child in the two continents most affected by desertification.

The UN Convention to Combat Desertification, 1994

The most recent and meaningful chapter in international efforts to address the problem of desertification began in Rio de Janeiro in 1992. Starting with the UN Conference on the Human Environment held in Stockholm in 1972, each decade it has become a tradition for heads of state and senior ministers of the world to convene and discuss the state of the planet. The 1982 gathering in Nairobi was not nearly as dramatic in its results as the Stockholm conference. So when it came time to plan the 1992 event to be held in Rio de Janeiro, the United Nations was looking for global environmental issues that would galvanize the international community. Planning for the conference took place soon after publication of the UN report *Our Common Future* in which the Norwegian prime minister, Geo Brundtland, and her committee made "sustainable development" a modern shibboleth and international environmental paradigm.[27] There was a call to make the conceptual slogans more operational. A broad strategic document containing principles and objectives for global "sustainability" was prepared for approval of the conference, entitled *Agenda 21*. Although *Agenda 21* had a chapter on desertification, it was the issues of climate change and biodiversity loss that were the focus of the developed nations, who had grown alarmed at new data and negative trends.

Not everyone was thrilled with the orientation of the conference planners. In particular, many African countries were less than enthusiastic about the agenda. They argued that if global warming existed, it was caused by the excess emissions of wealthy countries that should solve the problem themselves. As to biodiversity loss, while important, surely it was of far less significance than the pervasive poverty that plagued the African continent. In short, given the desperate situation in Africa, what was really needed was a convention to address poverty and Africa.

The Western donor countries were wary of expanding existing commitments for direct poverty relief, but they saw a middle ground in desertification—an environmental issue that was a key driver of global poverty. When the UN General Assembly passed its formal resolution convening the UN Conference on Environment and Development, it called "to accord the issue of desertification high priority." In retrospect, however, the reference appears to have been an exercise in lip service. During the Preparatory Committee meetings leading up to

Rio, desertification was largely absent from deliberations. When ministers from African states convened in November 1991 in Cote d'Ivoire, they were determined to lobby the UN to get serious about the subject. The ministers called for Rio's *Agenda 21* to include a recommendation to prepare a convention to combat desertification.

But at Rio itself, efforts to support a convention sputtered. The European Union blocked the idea, arguing that desertification by its nature was a regional issue and therefore fundamentally inappropriate for a global agreement. Ultimately, the United States changed its unsympathetic position and came to support the African demands. A compromise was reached: the conference called on the UN General Assembly to establish an Intergovernmental Negotiation Committee to start preparing a convention. The General Assembly assented and the committee was established, starting its work in Nairobi in May 1993 and finishing the draft in Paris. The agreement was ready for signatures by October 15, 1994. The document—known as United Nations Convention to Combat Desertification (UNCCD) in Those Countries Experiencing Serious Drought and/or Desertification, Particularly in Africa—required ratification by 50 countries to enter into force. By December 26, 1996, the requisite number of nations was on board. Today more than 190 nations are signatories.

The convention hardly offered a paragon of global governance or efficacious regulation of the commons, however. It established the usual institutions for an international agreement (a Convention Secretariat based in Bonn, a Committee to Review Implementation of the Convention, and a Committee on Science and Technology to advise the Conference of the Parties that meets every other year). Substantively, the convention divides the world into "haves" and "have-nots." Endorsing a "bottom-up" orientation, developing "affected nations" (the have-nots) are expected to prepare "action programs" in consultation with local communities. The associated projects are to be funded by donor countries in bilateral or multilateral partnerships with affected developing countries. The convention recognizes that there are indirect, social drivers of desertification that need to be addressed in National Action Plans. It also stipulates that legislation should be part of national strategies.

Progress in implementation has been painfully slow. Only in 2005 did most African nations complete a National Action Plan, and these are often extremely vague and short on specifics. Oversight by the Convention Secretariat is limited by its annual budget, currently a paltry eight million dollars after being cut by 30 percent at the 2006 Conference of the Parties in Nairobi.[28] Most important, the success of the convention is predicated on generous financial assistance from developed nations, and the first world has not ponied up. Although there is no shortage of local success stories that are the result of UNCCD commitments, no empirical evidence as yet points to a reduction in desertified lands on a global scale.

Yet the UNCCD is a dynamic agreement—still young and reasonably nimble. For example, recently the UNCCD adopted a new ten-year strategy that is based on "results based management" and quantifiable monitoring mechanisms.[29] There are sufficient signs to justify modest optimism and classification of the treaty as a major turning point. The Global Environmental Facility—the financial mechanism that awards grants for activities that protect biodiversity and reduce greenhouse gasses—recently decided to expand its mandate and include hundreds of dollars each year to support projects that address land degradation. There are literally hundreds of projects on the ground that are part of UNCCD related efforts. Many of them are run by local NGOs that are flourishing throughout Africa, Asia, and South America. Desertification is deemed a lucrative enough field for international aid to motivate many countries without a centimeter of drylands in them, to seek international support to fight their desertification problem—and even to prepare national action programs.

Mapping the Scope of Desertification

An essential element for a global response to desertification involves monitoring progress. Without clear benchmarks of land and vegetative conditions, it is impossible to evaluate whether international efforts are successful or not. There have been at least four attempts to date to assess the scope and the severity of desertification on the planet. None are perfect, but there seems to be a steady improvement in the resolution. When the Millennium Ecosystem Assessment prepared its synthesis report in 2005, it could speak with some confidence about the scope of the problem. Initially information was collected for the UNCOD gathering in 1977, and the same database provided the basis for the World Map of Desertification that was prepared by the UN's Food and Agriculture Organization, UNESCO, and the World Meteorological Organization soon thereafter. But this map only showed areas that were vulnerable to desertification and provided little information about actual conditions on the ground. It was a start, though.

A few years later the soil scientist Harold Dregne, working with a team from Texas Tech University, looked at the state of soil and vegetation degradation in some one hundred countries. The results were more satisfying methodologically but far from perfect. The authors relied on anecdotal accounts, research reports, travelers' descriptions, personal opinions, and local experience. From a substantive point of view, however, the results were downright alarming, with an aggregate estimate that 70 percent of the world's drylands suffered from desertification. In 1991 the UN Environmental Program commissioned the Global Assessment of Soil Degradation (GLASOD). Prepared in the Netherlands, the data relied on questionnaires circulated to soil experts around the world, and included information about the type, degree, extent, cause, and rate of soil degradation. The report did not consider the so-called hyperarid regions, but of the remaining drylands, it suggested that a fifth suffered from human-induced soil degradation. This left experts with the impression that Dregne's 1983 study had grossly overstated the severity of the situation.

In 2003 the Partial Coverage Assessment was produced at the behest of the Millennium Assessment project. This initiative produced a "desk study" that built on past efforts. It focused on soil conditions (rather than vegetation degradation) and was based on existing literature, erosion models, regional data sets, and remote sensing. Here, the numbers were even less severe. Only about 10 percent of the world's drylands were categorized as degraded. After reviewing the sundry mapping exercises, the Millennium Assessment (the closest thing to a scientific consensus estimate ever compiled) felt comfortable with a 10 percent to 20 percent range of global dryland degradation—still a staggering amount of damaged territory, given that 47 percent of the earth's continental surface is drylands.[30] It is somewhat discouraging to think that a full thirty years after the United Nations had turned its attention to the problem of desertification, only today are clear baseline numbers beginning to coalesce by which progress in the field can be measured. Yet, assuming that efforts to address land degradation escalate, it is truly a case of better late than never. Only when the planet knows where it stands, can it hope to figure out where it should be going.

China Legislates a Great Wall to Combat Desertification, 2001

China is an example of a nation that has decided that it can no longer take a "business as usual" approach to the problem of desertification. The crisis is simply too severe, threatening the country with economic collapse. Environmental icon Lester Brown has described the grimness of the situation: "China's Environmental Protection Agency reports that the Gobi Desert expanded by 52,400 square kilometers (20,240 square miles) from 1994 to 1999, an area half the size of Pennsylvania. With the advancing Gobi now within 150 miles of Beijing, China's leaders are beginning to sense the gravity of the situation. With little vegetation remaining in parts of northern and western China, the strong winds of late winter and early spring can remove literally millions of tons of topsoil in a single day—soil that can take centuries to replace. On April 12, 2002, South Korea was engulfed by a huge dust storm from China that left people in Seoul literally gasping for breath."[31]

In 1996, China was among the first nations to draft a national action program and it was an ambitious one. The Chinese program envisions complete rehabilitation of local ecosystems. By the year 2050, the program says, "nearly all desertified land will be brought into control."[32] This initially was to be achieved through the expansion of a range of afforestation and agricultural projects to stem the tide of desertification. For example, the Sanbei Shelterbelt was already well under way. Begun in 1978, the windbreak forest was soon declared the "planet's largest ecological project." Billions of work days by Chinese peasants have gone into the initial phase, and the tree line will eventually stretch along a 1.3-million-kilometer (780,000 miles) line.[33]

But mega projects alone are not enough. Fundamental changes in land use and agriculture are required. In 2001, China passed legislation that transformed the operational aspects of its national action program into an ambitious antidesertification law.[34] The law's regulatory orientation is unprecedented. The Law of Desertification Prevention and Control of the People's Republic of China *drastically* limits grazing and cultivation on vulnerable lands. It is unlawful to commit an act of vegetation destruction, and the government has the authority to issue orders to stop erosive activities. Any profits from activities that contribute to desertification are illegal and can be confiscated.[35] If a government agency violates this law by failing to report land deterioration, failing to erect sand and windbreaks, approving cultivated land in vegetative areas, and developing and constructing on unapproved areas, the overseeing administrative body will be sanctioned.[36] Along with the stick there is a carrot: the law establishes financial incentives for farmers who rehabilitate degraded land.[37] Individuals who rehabilitate land or set it aside as a protected area or natural reserve are to be compensated.

It is not yet clear whether China will succeed in turning the tide in its battle with desertification. It is also much clearer that most nations do not have a political culture that will allow them to promulgate such stringent command-and-control policies for land use. Yet the Chinese legislation and its implementation offer a model that may offer inspiration about the intensity that human efforts to successfully stop land degradation can take. Of course it can be argued that what China calls desertification is largely the natural movement of sand dunes. Or some might maintain that China's efforts are local in nature and do not represent a global turning point. To this it can be said that China remains the most populated country in the world, and the unique magnitude of its crisis, and the comprehensive single-minded national response, warrants "turning point" designation.

The International Year of Deserts and Desertification, 2006

When the UN Environmental Program Governing Council convened in 2005, it called on the UN General Assembly to give the desertification issue the push forward it had long needed. UNEP "invited the General Assembly to consider declaring an International Year of Deserts and Desertification."[38] It would take less than a year for the General Assembly to respond and pass a resolution to that end. The "IYDD" began with little funding and only a little more fanfare. Nations were expected to launch events that would raise awareness about the problem of desertification and galvanize decision makers and the general public to become engaged. At the same time the UNCCD secretariat emphasized that the year was also an opportunity to remind the world that deserts contain unique and beautiful ecosystems and were the homes of splendid ancient civilizations: "They stand like open-air museums, bearing witness to bygone eras. The Year will therefore also celebrate the fragile beauty and unique heritage of the world's deserts, which deserve protection."[39]

Dozens of international conferences, workshops, and film festivals were quickly organized to mark the year, from Beijing to Rome to Israel's Negev desert. On the domestic front many countries embraced the UN's "call to arms." For example, by July 2006, Portugal could report more than ninety IYDD-related events (conferences, exhibitions, and so on), ten television programs dedicated to the topic, parliamentary hearings, and over three hundred articles in the written press, exposing thousands of citizens to the issue.[40] Looking back, though, it is doubtful whether the International Year of Deserts and Desertification will appear as a "turning point" any more than the events summarized in this chapter were truly pivotal watersheds that revolutionized attitudes and solved an acute environmental problem of global dimensions. Indeed, the events highlighted here can more precisely be seen as milestones in a slow and continuous voyage that suffers from the usual sluggishness of international diplomacy and, more important, the general lack of funds that characterize the international response to Africa and the developing world's distress.

Given the high population densities in many affected countries, solving the problem of desertification will require the creation of alternative livelihoods, enabling people to ease their dependence on the beleaguered soil to earn a living. Reducing birthrates and family planning will have to be elevated and become paramount policy objectives in most developing dryland countries suffering desertification. The road to such sustainable development is a painfully slow and sometimes mundane crawl, rather than an earth-shattering quantum leap or turning point. It is of some comfort to know that the international community has finally begun to join the thousands of dryland communities that suffer from desertification around the world on this journey. But given the enormity of the problem, it is not enough.

Notes

Chapter 1. Thinking Big

1. Carolyn Merchant, *The Columbia Guide to American Environmental History* (New York: Columbia University Press, 2002), 206.

2. Lynn White Jr., "The Historical Roots of Our Ecologic Crisis," *Science* 155 (1967): 1203–7.

3. Rolf Peter Sieferle, *The Subterranean Forest: Energy Systems and the Industrial Revolution* (Cambridge: White Horse Press, 2001).

4. See, for instance, Günther E. Thüry, *Die Wurzeln unserer Umweltkrise und die griechisch-römische Antike* (Salzburg: O. Müller, 1995); and Günter Bayerl, "Prolegomenon der 'Großen Industrie,'" in *Umweltgeschichte. Umweltverträgliches Wirtschaften in historischer Perspektive,* Acht Beiträge, edited by Werner Abelshauser (GG Sonderheft 15, Gottingen: Vandenhoeck Ruprecht, 1994), 29–56.

5. See also the ongoing controversy stimulated by Joachim Radkau; see his "Holzverknappung und Krisenbewußtsein im 18. Jahrhundert," *Geschichte und Gesellschaft* 9 (1983): 513–43.

6. John R. McNeill, "Observations on the Nature and Culture of Environmental History," *History and Theory* 42 (2003): 9.

7. Ibid., 6.

8. See also Sidney Pollard, "Industrialization and the European Economy," *Economic History Review* 26 (1973): 636–48.

9. On the concept of vulnerability and its importance for studies of natural disasters, see Susanna M. Hoffman and Anthony Oliver-Smith, eds., *Catastrophe and Culture: The Anthropology of Disaster* (Oxford: J. Currey, 2002).

10. Alfred W. Crosby, *The Columbian Exchange: Biological and Cultural Consequences of 1492* (Westport, Conn.: Greenwood, 1972).

11. William Cronon, *Nature's Metropolis: Chicago and the Great West* (New York: W. W. Norton, 1992).

12. Charles S. Maier, "Consigning the Twentieth Century to History: Alternative Narratives for the Modern Era," *American Historical Review* 105 (2000): 807–31.

13. See also Frank Uekoetter, *The Green and the Brown: A History of Conservation in Nazi Germany* (New York: Cambridge University Press, 2006).

14. John R. McNeill, *Something New under the Sun: An Environmental History of the Twentieth Century* (London: Allen Lane, 2000), xx.

15. Will Steffen, Paul J. Crutzen, and John R. McNeill, "The Anthropocene: Are Humans Now Overwhelming the Great Forces of Nature?" *Ambio* 36, no. 8 (2007): 614–21.

16. Fernand Braudel, *The Mediterranean and the Mediterranean World in the Age of Philip II*, vol. 1 (London: Collins, 1972), 21.

17. Kendall E. Bailes, "Critical Issues in Environmental History," in *Environmental History: Critical Issues in Comparative Perspective*, edited by Kendall Bailes (Lanham, Md.: University Press of America, 1985), 10.

Chapter 2. The First Hundred Thousand Years

1. Johan Goudsblom, *Fire and Civilization* (London: Allen Lane, 1992).

2. The capacity of humans and horses to work together, and the ability for horse and rider to cover long distances quickly, depends on this shared trait of profuse sweating.

3. See Tim Flannery, *The Future Eaters* (New York: Braziller, 1994), for a discussion of the late Pleistocene overkill hypothesis as it applies to Australia.

4. For a recent overview emphasizing climate's role, see Arlene Miller Rosen, *Civilizing Climates: Social Responses to Climate Change in the Ancient Near East* (Lanham, Md.: AltaMira Press, 2007).

5. Peter Bellwood, *The First Farmers: Origins of Agricultural Societies* (Oxford: Blackwell, 2005); and Melinda Zeder, Daniel Bradley, Eve Emshwiller, and Bruce Smith, eds., *Documenting Domestication: New Genetic and Archeological Paradigms* (Berkeley: University of California Press, 2006).

6. These matters are reviewed in Jared Diamond, *Guns, Germs, and Steel* (New York: Norton, 1997).

7. On the genetic evolution of lactose tolerance, see Gregory Cochran and Henry Harpending, *The Ten Thousand Year Explosion: How Civilization Accelerated Human Evolution* (New York: Basic Books, 2009), 77–78, 83–84.

8. James Webb, *Humanity's Burden: A World History of Malaria* (New York: Cambridge University Press, 2008).

9. David H. Thomas, *Native American Landscapes of St. Catherines Island, Georgia* (Washington, D.C.: American Museum of Natural History, 2008), 3 vols.

10. Mark Nathan Cohen, *Health and the Rise of Civilization* (New Haven, Conn.: Yale University Press, 1989); and Jean-Pierre Bocquet-Appel and Stephan Naji, "Testing the Hypothesis of a Worldwide Neolithic Demographic Transition," *Current Anthropology* 47 (2006): 341–65.

11. Llama and alpaca herders of the pre-Columbian Andes were generally not pastoralists but agriculturalists who raised animals as a sideline.

12. Kautilya, *The Arthashastra* (London: Penguin Classics, 1962), 2.2.6–7, cited in Sunil Sen Sarma, "Contemporaneity of the Perception of Environment in Kautilya's Arthasastra," *Indian Journal of History of Science* 33 (1998): 37–50.

13. Alan Macfarlane, *The Savage Wars of Peace: England, Japan, and the Malthusian Trap* (Oxford: Blackwell, 1997), 22.

14. Vaclav Smil, *Energies* (Cambridge: MIT Press, 1999).

15. J. R. McNeill and Verena Winiwarter, "Soils, Soil Knowledge, and Environmental History," in *Soils and Societies: Perspectives from Environmental History,* edited by J. R. McNeill and Verena Winiwarter (Cambridge: White Horse Press, 2005), 1–3.

16. Anzelm Gostomski, *Gospodarstwo* (1588; reprint, Wydawn, Poland: Zakładu Narodowego im. Ossolińskich, 1951).

17. William Ruddiman, *Plows, Plagues, and Petroleum: How Humans Took Control of Climate* (Princeton, N.J.: Princeton University Press, 2005), 65–114.

18. Michael Williams, *Deforesting the Earth: From Prehistory to Global Crisis* (Chicago: University of Chicago Press, 2003).

19. Peter Perdue, *China Marches West: The Qing Conquest of Central Asia* (Cambridge: Harvard University Press, 2005).

20. John Iliffe, *Africans: The History of a Continent* (Cambridge: Cambridge University Press, 1995); and Mark Elvin, *The Retreat of the Elephants: An Environmental History of China* (New Haven, Conn.: Yale University Press, 2004).

21. J. R. McNeill, "Biological Exchange and Biological Invasion in World History," in *Making Sense of Global History*, edited by Sølvi Sogner (Oslo: Universitersforlaget, 2001), 106–18.

22. Alfred Crosby, *The Columbian Exchange: The Biological and Cultural Consequences of 1492* (Westport, Conn.: Greenwood, 1972).

23. It is uncertain, although in my view likely, that the Black Death affected parts of Africa.

24. Crosby, *Columbian Exchange*; and Judith Carney, *Black Rice: The African Origins of Rice Cultivation in the Americas* (Cambridge: Harvard University Press, 2001).

25. John F. Richards, *The Unending Frontier: An Environmental History of the Early Modern World* (Berkeley: University of California Press, 2003); Warren Dean, *With Broadax and Firebrand: The Destruction of the Brazilian Atlantic Forest* (Berkeley: University of California Press, 1995); and Reinaldo Funes Monzote, *De bosque a sabana: Azúcar, deforestación y medio ambiente en Cuba: 1492–1926* (Mexico City: Siglo XXI, 2004).

26. Peter Thorsheim, *Inventing Pollution: Coal, Smoke, and Culture in Britain since 1800* (Athens: Ohio University Press, 2006).

27. Comment on the effects of fossil fuels may be found in J. R. McNeill, *Something New under the Sun: An Environmental History of the Twentieth-century World* (New York: Norton, 2000), and in any number of works by Vaclav Smil: see Smil, *Energy in World History* (Boulder, Colo.: Westview, 1994); Smil, *Energies: An Illustrated Guide to the Biosphere and Civilization* (Cambridge: MIT Press, 1999); Smil, *Energy in Nature and Society: General Energetics of Complex Systems* (Cambridge: MIT Press, 2008).

Chapter 3. Agriculture

1. See, for example, Joan Thirsk, *England's Agricultural Regions and Agrarian History, 1500–1750* (London: Economic History Society, 1987); Sidney Mintz, *Sweetness and Power: The Place of Sugar in Modern History* (New York: Viking, 1985); Elinor Melville, *A Plague of Sheep: Environmental Consequences of the Conquest of Mexico* (Cambridge: Cambridge University Press, 1994); and Donald Worster, *Dust Bowl: The Southern Plains in the 1930s* (New York: Oxford University Press, 1979).

2. J. D. Chambers and G. E. Mingay, *The Agricultural Revolution, 1750–1880* (New York: Schocken, 1966); and E. Kerridge, *The Agricultural Revolution* (London: Allan and Unwin, 1967).

3. See, for example, Pete Daniel, *Breaking the Land: The Transformation of Cotton, Tobacco, and Rice Cultures since 1880* (Urbana: University of Illinois Press, 1985); and Joyce Chaplain, *An Anxious Pursuit: Agricultural Innovation and Modernity in the Lower South, 1730–1815* (Chapel Hill: University of North Carolina Press, 1993). For the plantations, see Richard Drayton, *Nature's Government: Science, Imperial Britain, and the "Improvement" of the World* (New Haven, Conn.: Yale University Press, 2000).

4. James Scott, *Seeing Like a State: How Certain Schemes to Improve the Human Condition Have Failed* (New Haven, Conn.: Yale University Press, 1998).

5. Margaret Rossiter, *The Emergence of Agricultural Science: Justus Liebig and the Americans, 1840–1880* (New Haven, Conn.: Yale University Press, 1975). See also Steven Stoll, *Larding the Lean Earth: Soil and Society in Nineteenth-century America* (New York: Hill and Wang, 2002); and Frank Uekoetter, "Know Your Soil: Transitions in Farmers' and Scientists' Knowledge in the Twentieth Century," in *Soils and Socieeties: Perspectives from Environmental History*, edited by John McNeill and Verena Winiwarter, 322–40 (Isle of Harris: White Horse Press, 2006).

6. A. Hunter Dupree, *Science in the Federal Government: A History of Policies and Activities* (Cambridge: Belknap Press of Harvard University Press, 1957); Charles Rosenberg, *No Other Gods: On Science and American Social Thought* (Baltimore, Md.: Johns Hopkins University Press, 1976); and David Danbom, *Born in the Country: A History of Rural America* (Baltimore, Md.: Johns Hopkins University Press, 1995).

7. David Danbom, *The Resisted Revolution: Urban America and the Industrialization of Agriculture, 1900–1930* (Ames: Iowa State University Press, 1979); and Mary Neth, *Preserving the Family Farm: Women, Community, and the Foundations of Agribusiness in the Midwest, 1900–1940* (Baltimore, Md.: Johns Hopkins University Press, 1995).

8. William Cronon, *Nature's Metropolis: Chicago and the Great West* (New York: W. W. Norton, 1991); Ronald Kline, *Consumers in the Country: Technology and Social Change in Rural America* (Baltimore, Md.: Johns Hopkins University Press, 2000); Hal Barron, *Mixed Harvest: The Second Great Transformation in the Rural North, 1870–1930* (Chapel Hill: University of North Carolina Press, 1997); and Steven Stoll, *The Fruits of Natural Advantage: Making the Industrial Countryside in California* (Berkeley: University of California Press, 1998).

9. Willard Cochrane, *The Development of American Agriculture: A Historical Analysis* (Minneapolis: University of Minnesota Press, 1993); Deborah Fitzgerald, *Every Farm a Factory: The Industrial Ideal in American Agriculture* (New Haven, Conn.: Yale University Press, 2003); and Kline, *Consumers in the Country*.

10. Worster, *Dust Bowl*; and Fitzgerald, *Every Farm a Factory*.

11. Pare Lorentz's films *The River* and *The Plow That Broke the Plains* are classic propaganda films generated by the New Deal agencies eager to persuade farmers to change their field practices. See also Sara Gregg, "Uncovering the Subsistence Economy in the Twentieth-century South: Blue Ridge Mountain Farms," *Agricultural History* 78 (Fall 2004); and Randal S. Beeman and James A. Pritchard, *A Green and Permanent Land: Ecology and Agriculture in the Twentieth Century* (Lawrence: University Press of Kansas, 2001).

12. Deborah Fitzgerald, *The Business of Breeding: Hybrid Corn in Illinois, 1890–1940* (Ithaca, N.Y.: Cornell University Press, 1990); and Lawrence Busch, William B. Lacy, Jeffrey Burkhardt, and Laura Lacy, *Plants, Power, and Profit: Social, Economic, and Ethical Consequences of the New Biotechnologies* (Cambridge: Blackwell, 1991).

13. Harvey Levenstein, *Revolution at the Table: The Transformation of the American Diet* (New York: Oxford, 1988); Harvey Levenstein, *Paradox of Plenty: A Social History of Eating in Modern America* (New York: Oxford, 2003); and "Report of the Field Trip of Dr. Irwin C. Feustel," January 23, 1940 (National Archives, [San Bruno, Calif.], Western Regional Research Laboratory, USDA, Agricultural Research Service, RG 310, 90, 002, box 4).

14. The Iowa story is based on Deborah Fitzgerald, "Eating and Remembering," *Agricultural History* 79 (Fall 2005).

15. This is described in Steve Striffler, *Chicken: The Dangerous Transformation of America's Favorite Food* (New Haven, Conn.: Yale University Press, 2005). Shane Hamilton, "Trucking Country: Food Politics and the Transformation of Rural Life in Postwar America" (Ph.D diss., Massachusetts Institute of Technology, 2005).

16. For a thoughtful look at a particular region, see Jane Adams, *The Transformation of Rural Life: Southern Illinois, 1890–1990* (Chapel Hill: University of North Carolina Press, 1994).

17. Fitzgerald, "Eating and Remembering." For other agricultural chemicals, see Pete Daniel, *Toxic Drift: Pesticides and Health in the Post–World War II South* (Washington, D.C.: Smithsonian Institution Press, 2005).

18. Mark Kramer, *Three Farms: Making Meat, Milk, and Money from the American Soil* (Boston: Little, Brown, 1980).

19. Although historians have been somewhat slow to enter this discussion, journalists have been energetic; see, for example, Daniel Charles, *Lords of the Harvest: Biotech, Big Money, and the Future of Food* (New York: Perseus, 2001); and Bill Lambrecht, *Dinner at the New Gene Café: How Genetic Engineering Is Changing What We Eat, How We Live, and the Global Politics of Food* (New York: Thomas Dunne Books, 2001).

20. Historians have not yet addressed the international dimensions very much. But see, for example, Akhil Gupta, *Postcolonial Developments: Agriculture in the Making of Modern India* (Durham, N.C.: Duke University Press, 1998).

21. Thirsk, *England's Agricultural Regions and Agrarian History*.

22. John McNeill, *Something New under the Sun: An Environmental History of the Twentieth-century World* (New York: Norton, 2000), offers a good model for this kind of project.

Chapter 4. Forest History

1. See Peter Blickle, *Der Bauernkrieg: Die Revolution des gemeinen Mannes* (Munich: Beck, 1998).

2. On the Hambacker Fest (the Hambach Celebration), see Kultusministerium Rheinland-Pfalz, ed., *Hambacher Fest, 1832: Freiheit und Einheit, Deutschland und Europa. Katalog zur Dauerausstellung* (Neustadt: Meininger, 1990).

3. For more about this region's forests, see Bernd-Stefan Grewe, *Der versperrte Wald: Ressourcenmangel in der bayerischen Pfalz (1814–1870)* (Cologne: Boehlau, 2004).

4. This parallel was intended by the organizers when they chose the same date and place as in 1525.

5. See also Rolf Peter Sieferle, "Die Grenzen der Umweltgeschichte," *GAIA* 2 (1993): 8–21.

6. See Erich Bauer, *Los montes de España en la Historia* (Madrid: Minister of Agriculture, 1980).

7. Christof Dipper, *Deutsche Geschichte, 1648–1789* (Frankfurt: Suhrkamp, 1991), 41; and Uwe Eduard Schmidt, *Der Wald in Deutschland im 18. und 19. Jahrhundert: Das Problem der Ressourcenknappheit* (Saarbrücken: Conte, 2002).

8. Dorothea Hauff, *Zur Geschichte der Forstgesetzgebung und Forstorganisation des Herzogtums Württemberg im 16. Jahrhundert* (Stuttgart: Landesforstverwaltung Baden-Wuerttemberg, 1977); Joachim Allmann, *Der Wald in der frühen Neuzeit: Eine mentalitäts- und sozialgeschichtliche Untersuchung am Beispiel des Pfälzer Raumes, 1500–1800* (Berlin: Duneker & Humblot, 1989), 346; and Ingrid Schäfer, *Ein Gespenst geht um Politik mit der Holznot in Lippe, 1750–1850. Eine Regionalstudie zur Wald- und Technikgeschichte* (Detmold: NHV Lippe, 1992), 230.

9. On financial politics, see Hanna Schissler, "Preußische Finanzpolitik nach 1807: Die Bedeutung der Staatsverschuldung als Faktor der Modernisierung des preußischen Finanzsystems," *Geschichte und Gesellschaft* 8 (1982):367–85; Hans-Peter Ullmann, *Staatsschulden und Reformpolitik: Die Entstehung moderner öffentlicher Schulden in Bayern und Baden, 1780–1820*, 2 vols. (Göttingen: Vandenhoeck & Ruprecht, 1986); Eckart Schremmer, *Steuern und Staatsfinanzen während der Industrialisierung Europas: England, Frankreich, Preußen, und das Deutsche Reich 1800 bis 1914* (Berlin: Springer, 1994); and Werner Buchholz, *Geschichte der öffentlichen Finanzen in Europa in Spätmittelalter und Neuzeit: Darstellung, Analyse, Bibliographie* (Berlin: Akademie, 1996).

10. See Hans-Walter Keweloh, ed., *Auf den Spuren der Flößer: Wirtschafts- und Sozialgeschichte eines Gewerbes* (Stuttgart: Konrad Theiss, 1988); Dietrich Ebeling, *Der Holländerholzhandel in den Rheinlanden: Zu den Handelsbeziehungen zwischen den Niederlanden und dem westlichen Deutschland im 17. und 18. Jahrhundert* (Stuttgart: Steiner, 1992); Max Scheifele, *Als die Wälder auf Reisen gingen: Wald – Holz – Flößerei in der Wirtschaftsgeschichte des Enz-Nagold-Gebietes* (Karlsruhe: Braun, 1996).

11. See August Bernhardt, *Geschichte des Waldeigentums, der Waldwirtschaft und Forstwissenschaft in Deutschland*, 3 vols. (Berlin: J. Springer, 1872–1875).

12. See Stefan Brakensiek, *Agrarreform und ländliche Gesellschaft: Die Privatisierung der Marken in Nordwestdeutschland 1750–1850* (Paderborn: Schoeningh, 1991); Marie-Danielle Demélas and Nadine Vivier, eds., *Les propriétés collectives face aux attaques libérales (1750–1914): Europe occidentale et Amérique latine* (Rennes: Presses Universitaires de Rennes, 2003).

13. For England, France, and Germany, see Heinrich Rubner, *Forstgeschichte im Zeitalter der Industriellen Revolution* (Berlin: Duncker & Humblot, 1967).

14. Werner Sombart, *Der moderne Kapitalismus: Historische systematische Darstellung des gesamteuropäischen Wirtschaftslebens von seinen Anfängen bis zur Gegenwart*, vol. 2, *Das europäische Wirtschaftsleben im Zeitalter des Frühkapitalismus*, 3 vols. (Berlin: Duncker & Humblot, 1969) (first edition 1916), 1114.

15. On the importance of the forests for the peasants' economy, see Bernward Selter, *Waldnutzung und ländliche Gesellschaft: Landwirtschaftlicher "Nährwald" und neue Holzökonomie im Sauerland des 18. und 19. Jahrhunderts* (Paderborn: Schoeningh, 1995); and Winfried Schenk, *Waldnutzung, Waldzustand und regionale Entwicklung in vorindustrieller Zeit im mittleren Deutschland: Historisch-geographische Beiträge zur Erforschung von Kulturlandschaften in Mainfranken und Nordhessen* (Stuttgart: Steiner, 1996).

16. See Rainer Fremdling, *Technologischer Wandel und internationaler Handel im 18. und 19. Jahrhundert: Die Eisenindustrien in Großbritannien, Belgien, Frankreich, und Deutschland* (Berlin: Duncker & Humblot, 1986); and Akos Paulinyi, "Die Umwälzung der Technik in den Industriellen Revolution zwischen 1750 und 1840," in *Mechanisierung und Maschinisierung, 1600 bis 1840*, edited by Ulrich Troitzsch (Berlin: Ullstein, 1991), 271–498.

17. Max Endres, *Handbuch der Forstpolitik mit besonderer Berücksichtigung der Gesetzgebung und Statistik*, second edition (Berlin: J. Springer, 1922), 586.

18. On the global wood trade, see Richard Tucker and J. F. Richards, eds., *Global Deforestation and the Nineteenth-century World Economy* (Durham, N.C.: Duke University Press, 1983).

19. Max Endres, *Die Waldbenutzung, Vom 13. bis Ende des 18. Jahrhunderts: Ein Beitrag zur Geschichte der Forstpolitik* (Tübingen: Laupp, 1888).

20. On offenses against the forest orders and forest laws, see Josef Mooser, "'Furcht bewahrt das Holz': Holzdiebstahl und sozialer Konflikt in der ländlichen Gesellschaft 1800–1850 an westfälischen Beispielen," in *Räuber, Volk, und Obrigkeit,* edited by Heinz Reif

(Frankfurt: Suhrlamp, 1984), 43–99; Reiner Prass, "Verbotenes Weiden und Holzdiebstahl: Ländliche Forstfrevel am südlichen Harzrand im späten 18. und frühen 19. Jahrhundert," *Archiv für Sozialgeschichte* 36 (1996): 51–68; and Bernd-Stefan Grewe, "Forstfrevel—(k)ein Bagatelldelikt," in *Unrecht und Recht: Kriminalität und Gesellschaft im Wandel von 1500–2000*, edited by Heinz-Günther Borck and Beate Dorfey (Koblenz: Landeshauptarchiv, 2002), 527–42.

21. On professionalization, see Thomas Götz, "Der Staat im Wald: Forstpersonal und Forstausbildung in Bayern im ersten Drittel des 19. Jahrhunderts aus umweltgeschichtlicher Perspektive: Das Beispiel des Rheinkreises," in *Beiträge zur Umweltgeschichte I*, edited by Christoph Ernst, Bernd-Stefan Grewe, and Joachim Kuntz (Trier: Universität Trier, 1996), 36–77. In other countries this happened later in the mid-nineteenth century: in Spain with the Escuela Técnica Especial de Montes en Villaviciosa de Odón/Madrid 1847 and the creation of the Cuerpo de Ingenieros de Montes 1854 (Bauer, Los Montes); and in Russia in 1840. See Walter Kremser, *Epochen der Forstgeschichte Estlands* (Tallin: Tallina Raamatutruekikoda, 1998), 248 and following.

22. For example, see Hans Wiesmeth, *Umweltökonomie: Theorie und Praxis im Gleichgewicht* (Berlin: Springer, 2003).

23. There is no contradiction to what has been said here. The rulers were motivated to extend their control over the forests and therefore used the argument of overused and degraded forests that really existed.

24. The debate about values is of high importance for environmental history. See Joachim Radkau, *Natur und Macht: Eine Weltgeschichte der Umwelt* (Munich: Beck, 2000), 33–41.

25. On the development of "Waldbau," see Kurt Mantel, *Wald und Forst in der Geschichte: Ein Lehr- und Handbuch* (Alfeld-Hannover: Schaper, 1990), 321–78.

26. See Matthias Bürgi, *Waldentwicklung im 19. und 20. Jahrhundert: Veränderungen in der Nutzung und Bewirtschaftung des Waldes und seiner Eigenschaften als Habitat am Beispiel der öffentlichen Waldungen im Züricher Unter- und Weinland* (Zürich: ETH Zürich, 1998); and Ute Fenkner-Voigtländer, *Forsteinrichtung und Waldbau im Elmsteiner Wald unter deutschen und französischen Einflüssen, 1760–1860: Ein Beitrag zur Forstgeschichte des Pfälzerwaldes* (Mainz: Ministerium für Landwirtschaft, Weinbau, und Forsteu, 1992).

27. Christian Pfister, ed., *Das 1950er Syndrom: Der Weg in die Konsumgesellschaft* (Bern: Paul Haupt, 1995).

Chapter 5. The Nation-State

1. See Donald Worster, "Appendix: Doing Environmental History," in *The Ends of the Earth: Perspectives on Modern Environmental History*, edited by Donald Worster (Cambridge: Cambridge University Press, 1988), 289–307, here 289–90; 303. More recent analyses of the field still maintain the boundaries between material and political or cultural

environmental histories; see J. R. McNeill, "Observations on the Nature and Culture of Environmental History," *History and Theory* 42 (December 2003): 5–43, especially 35.

2. Worster, "Appendix: Doing Environmental History," 301.

3. Charles Maier, "Consigning the Twentieth Century to History: Alternative Narratives of the Modern Era," *American Historical Review* 105, no. 3 (June 2000): paragraph 3. Available online at http://www.historycooperative.org/journals/ahr/105.3/ah000807.html (accessed on December 20, 2005).

4. James C. Scott, *Seeing Like a State: How Certain Schemes to Improve the Human Condition Have Failed* (New Haven, Conn.: Yale University Press, 1998).

5. Ibid., 8.

6. Ibid.

7. Maier, "Consigning the Twentieth Century to History," 2.

8. On the role of institutions in shaping environmental discourse, see Joachim Radkau, "Germany as a Focus of European 'Peculiarities' in Environmental History," in *Germany's Nature: Cultural Landscapes and Environmental History*, edited by Thomas Lekan and Thomas Zeller (New Brunswick, N.J.: Rutgers University Press, 2005), 18–19.

9. Among the numerous examples of environmental-historical engagement with Scott's work are Karl Jacoby, *Crimes against Nature: Squatters, Poachers, Thieves, and the Hidden History of American Conservation* (Cambridge: Cambridge University Press, 2001); Paul Sutter, "Reflections: What Can U.S. Environmental Historians Learn from Non-U.S. Environmental Historiography?" *Environmental History* 8, no. 1 (January 2003): 109–29; and Nancy Langston, "On Teaching World Forest History," *Environmental History* 10, no. 1 (January 2005): 20–29.

10. As Ted Steinberg has noted, all structures, including those of the state, are "at risk" from a host of factors, especially ecological ones. See Steinberg, "Down to Earth: Nature, Agency, and Power in History," *American Historical Review* 107, no. 3 (June 2002): 802–3.

11. David Blackbourn, "Time Is a Violent Torrent: Constructing and Reconstructing Rivers in Modern German History," in *Rivers in History: Perspectives on Waterways in Europe and North America*, edited by Christof Mauch and Thomas Zeller (Pittsburgh: University of Pittsburgh Press, 2008), 24.

12. Carolyn Merchant, *Ecological Revolutions: Nature, Gender, and Science in New England* (Chapel Hill: University of North Carolina Press, 1989).

13. I borrow the term "contact zones" from Mary Louise Pratt, *Imperial Eyes: Travel Writing and Transculturation* (London: Routledge, 1992).

14. On the timber shortage or *Holznot* debate, see Joachim Radkau and Ingrid Schäfer, *Holz: Ein Rohstoff in der Technikgeschichte* (Reinbek: Rowohlt, 1987); and Joachim Radkau, "Wood and Forestry in German History: In Quest of an Environmental Approach," *Environment and History* 2 (1996): 63–76. See also Henry Lowood, "The Calculating Forester: Quantification, Cameral Science, and the Emergence of Scientific Forestry Management in Germany," in *The Quantifying Spirit in the Eighteenth Century*, edited by T. Frangsmyr, J. J. Heilbron, and R. E. Rider (Berkeley: University of California Press, 1990), 315–42.

15. Langston, "On Teaching World Forest History," 23.

16. Ramachandra Guha, *The Unquiet Woods: Ecological Change and Peasant Resistance in the Himalaya* (Delhi: Oxford University Press, 1989); and Thaddeus Sunseri, "Forestry and the German Imperial Imagination: Conflicts over Forest Use in German East Africa," in *Germany's Nature*, edited by Lekan and Zeller, 81–110.

17. On these examples, see Gary B. Cohen and Franz A. G. Szabo, eds., *Embodiments of Power: Building Baroque Cities in Europe* (New York: Berghahn Books, 2008); William Cronon, *Changes in the Land: Indians, Colonists, and the Ecology of New England* (New York: Hill and Wang, 1983); Mark Cioc, *The Rhine: An Ecobiography, 1815–2000* (Seattle: University of Washington Press, 2002); and Mart Stewart, *"What Nature Suffers to Groe": Life, Labor, and Landscape on the Georgia Coast, 1680–1920* (Athens: University of Georgia Press, 1996).

18. Cioc, *Rhine*, 10. See also Thomas Lekan, "A 'Noble Prospect': Tourism, *Heimat*, and Conservation on the Rhine, 1880–1914," *Journal of Modern History* 81, no. 4 (December 2009): 824–58.

19. See Clarence Glacken, *Nature and Culture in Western Thought from Ancient Times to the End of the Eighteenth Century* (Berkeley: University of California Press, 1967); Donald Worster, *Nature's Economy: A History of Ecological Ideas* (Cambridge: Cambridge University Press, 1985); and Roderick Nash, *Wilderness and the American Mind* (New Haven, Conn.: Yale University Press, 1967).

20. On the development of the steamship industry, see Cioc, *Rhine*, 146–47; Lekan, "A 'Noble Prospect'"; and David Blackbourn, *The Conquest of Nature: Water, Landscape, and the Making of Modern Germany* (New York: W. W. Norton, 2006), 161–72.

21. Scott, *Seeing Like a State*, 88.

22. Ibid., 92.

23. A. Joshua West, "Forest and National Security: British and American Forestry Policy in the Wake of World War I," *Environmental History* 8, no. 2 (April 2003): 270–93; and Donald Worster, *Dust Bowl: The Southern Plains in the 1930s* (New York: Oxford University Press, 1979).

24. Scott, *Seeing Like a State*, 89.

25. Ibid., 193.

26. On such technologies, see Paul Josephson, *Industrialized Nature: Brute Force Technology and the Transformation of the Natural World* (Washington, D.C.: Island Press, 2002).

27. Ibid., 256.

28. Matthew N. Eisler, "Review of Paul R. Josephson, *Industrialized Nature: Brute Force Technology and the Transformation of the Natural World*," H-Environment, H-Net Reviews, June 2005. Available online at http://www.h-net.org/reviews/showrev.cgi?path=208301124312653 (accessed on January 15, 2005).

29. Mauch and Zeller, *Rivers in History*, 4.

30. Blackbourn, *Conquest of Nature*, 184–90.

31. Donald Worster, *Rivers of Empire: Water, Aridity, and the Growth of the American West* (New York: Pantheon, 1985).

32. John McNeill, *Something New under the Sun: An Environmental History of the Twentieth Century* (New York: W. W. Norton, 2000), 159.

33. Ibid., 159.

34. This information comes from "World's Fourth Largest Lake Nearly Dried Up," *CBSNews.com*. Available online at http://www.cbsnews.com/stories/2010/04/04/world/main6362456.shtml (accessed on April 26, 2010).

35. Ibid., 162; McNeill describes the environmental consequences of the Aral Sea project on pages 162–66.

36. Scott, *Seeing Like a State*, 193–97.

37. For an account of the Hetch Hetchy controversy, see Nash, *Wilderness and the American Mind*, 161–81; and John W. Simpson, *Dam! Water, Politics, Power, and Preservation in Hetch Hetchy and Yosemite National Park* (New York: Pantheon, 2005).

38. Cited in William H. Rollins, *A Greener Vision of Home: Cultural Politics and Environmental Reform in the German* Heimatschutz Movement, 1904–1918 (Ann Arbor: University of Michigan Press, 1997), 141. On Laufenburg, see also Ulrich Linse, *Von der Bittschrift zur Platzbesetzung: Konflikte um technische Grossprojekte: Laufenburg, Walchensee, Wyhl, Wackersdorf* (Berlin: J.H.W. Dietz, 1988).

39. On this point, see Blackbourn, *Conquest of Nature*, 189–250; and Thomas Lekan, *Imagining the Nation in Nature: Landscape Preservation and German Identity* (Cambridge: Harvard University Press, 2004), 204–51.

40. Lekan, *Imagining the Nation*, 19–73.

41. Sutter, "What Can U.S. Environmental Historians Learn?" 111–17. On conservation as the colonization of marginal lands and peoples, see also Steinberg, "Down to Earth," 815–18.

42. On the "moral ecology" of rural folk, see Jacoby, *Crimes against Nature*, 3.

43. Ibid., 14–15.

44. Marsh, *Man and Nature*, 29–30. Cited in Jacoby, *Crimes against Nature*, 15.

45. On this point, see Sunseri, "Forestry and the German Imperial Imagination."

46. Sutter, "What Can U.S. Environmental Historians Learn?" 114–15.

47. On this point, see Worster, *Nature's Economy*, 339–48; Colin Riordan, "Green Ideas in Germany: A Historical Survey," in *Green Thought in German Culture: Historical and Contemporary Perspectives*, edited by Colin Riordan (Cardiff: University of Wales Press), 27–28; and Raymond Dominick, *The Environmental Movement in Germany: Prophets and Pioneers, 1871–1971* (Bloomington: Indiana University Press, 1992), 141–42, 163–68.

48. See Thomas Lekan, "Saving the Rhine: Water, Ecology, and *Heimat* in Post–World War II Germany," in *Rivers in History*, edited by Mauch and Zeller, 110–36.

49. International Commission for the Protection of the Rhine Against Pollution, *Ecological Master Plan for the Rhine/"Salmon 2000"* (Koblenz: ICPR Technical-Scientific Secretary, 1990).

50. Cioc, *Rhine*, 186–92.

51. Cited in Mauch and Zeller, *Rivers in History*, 5. Annie Luo, "Three Gorges: A World So Changed," July 8, 2003, World Press.org. Available online at http://www.worldpress.org/Asia/1245.cfm#down (accessed on January 19, 2005).

52. Douglas McGray, "Pop-up Cities: China Builds a Bright Green Metropolis," *Wired* 19, no. 5 (2007). Available online at http://www.wired.com/wired/archive/15.05/feat_popup.html (accessed on June 27, 2009).

53. Fred Pearce, "Greenwash: The Dream of the First Eco-City Was Built on a Fiction," *The Guardian*, April 23, 2009. Available online at http://www.guardian.co.uk/environment/2009/apr/23/greenwash-dongtan-ecocity (accessed on June 27, 2009).

54. Jesse Fox, "What Sank Dongtan?" Sustainable Cities Blog, January 20, 2009. Available online at http://www.sustainablecityblog.com/2009/01/dongtan-modern-day-shangri-la/ (accessed on June 27, 2009).

55. Christina Larson, "China's Grand Plans for Eco-Cities Now Lie Abandoned," *Yale Environment 360* (April 6, 2009). Available online at http://e360.yale.edu/content/feature.msp?id=2138 (accessed on June 27, 2009).

56. McNeill, *Something New under the Sun*, 203–304.

Chapter 6. Urban Environmental History

1. See Martin V. Melosi, "The Place of the City in Environmental History," *Environmental History Review* 17 (Spring 1993): 1–23, reprinted in Martin V. Melosi, *Effluent America: Cities, Industry, Energy, and the Environment* (Pittsburgh: University of Pittsburgh Press, 2001); J. R. McNeill, "Observations on the Nature and Culture of Environmental History," *History and Theory* 42 (December 2003): 6–38; Christine Meisner Rosen and Joel A. Tarr, eds., "The Environment and the City," *Journal of Urban History* 20 (May 1994): 299–434; Samuel P. Hays, *Explorations in Environmental History* (Pittsburgh: University of Pittsburgh Press, 1998), 69–100; Joel A. Tarr, "Urban History and Environmental History in the United States: Complementary and Overlapping Fields," in *Environmental Problems in European Cities in the Nineteenth and Twentieth Century*, edited by Christoph Bernhnardt (New York: Munster, 2001), 25–40; Andrew C. Isenberg, "New Directions in Urban Environmental History," in *The Nature of Cities: Culture, Landscape, and Urban Space*, edited by Andrew C. Isenberg (Rochester, N.Y.: University of Rochester Press, 2006), xi–xix. Martin V. Melosi and Joel A. Tarr are currently editing a new series, *The History of the Urban Environment*, with the University of Pittsburgh Press.

2. The classic study of a city's penetration into its hinterland is William Cronon, *Nature's Metropolis: Chicago and the Great West* (New York: W. W. Norton, 1991). Cronon, however, does not deal with the city's quest for such substances as water and fuel.

3. Ian Douglas, *The Urban Environment* (London: Arnold, 1983).

4. Five international round tables have been held in Europe focusing on urban environmental history in the nineteenth and twentieth centuries. The published papers

from three of these conferences furnish an excellent introduction to the state of urban environmental history in Europe. See Christoph Bernhardt, ed., *Environmental Problems in European Cities in the Nineteenth and Twentieth Century* (New York: Waxman, 2001); Christoph Bernhardt and Geneviève Massard-Guilbaud, *The Modern Demon: Pollution in Urban and Industrial European Societies* (Clermont-Ferrand, France: Presses Universitaires Blaise-Pascal, 2002); and Dieter Schott, Bill Luckin, and Geneviève Massard-Guilbaud, eds., *Resources of the City: Contributions to an Environmental History of Modern Europe* (Burlington, Vt.: Ashgate Publishing Company, 2005). Conferences were also held in Siena in 2004, Paris in 2006, and Berlin in 2008. Some insightful comparative studies are Harold L. Platt, *Shock Cities: The Environmental Transformation and Reform of Manchester and Chicago* (Chicago: University of Chicago Press, 2005); Frank Uekoetter, "Divergent Responses to Identical Problems: Businessmen and the Smoke Nuisance in Germany and the United States, 1880–1917," *Business History Review* 73 (Winter 1999): 641–76; and Uekoetter, *The Age of Smoke: Environmental Policy in Germany and the United States, 1880–1970* (Pittsburgh: University of Pittsburgh Press, 2009). See also Dieter Schott, "Urban Environmental History: What Lessons Are There to Be Learnt?" *Boreal Environment Research* 9 (2004): 1–10.

5. All population figures for American cities cited in this chapter derive from U.S. Census, *Populations of the One Hundred Largest Cities and Other Urban Places in the United States, 1790–1990*, at http://www.census.gov/population/www/documentation/twps0027.html, with adjustments to 2000.

6. Joel A. Tarr, "The City and Technology," in *A Companion to American Technology*, edited by Carroll Pursell (New York: Blackwell Publishing, 2005), 98–102.

7. See "A Note on Urban Reactions to Natural Disasters," at the end of this chapter.

8. Anne W. Spirn, *The Granite Garden: Urban Nature and Human Design* (New York: Basic Books, 1984).

9. See, for example, Sarah S. Elkind, *Bay Cities and Water Politics: The Battle for Resources in Boston and Oakland* (Lawrence: University of Kansas Press, 1998).

10. The most recent survey of urban water supply can be found in Martin Melosi, *The Sanitary City: Urban Infrastructure in America from Colonial Times to the Present* (Baltimore, Md.: Johns Hopkins University Press, 2000). Still useful as an overview of the history of urban water supply is Nelson Blake, *Water for the Cities: A History of the Urban Water Supply Problem in the United States* (Syracuse, N.Y.: Syracuse University Press, 1956); an unusual perspective is provided in Matthew Gandy, *Concrete and Clay: Reworking Nature in New York City* (Cambridge: MIT Press, 2002), 19–51; and Michael Rawson, "The Nature of Water Reform and the Antebellum Crusade for Municipal Water in Boston," *Environmental History* 9 (July 2004): 411–35.

11. Melosi, *Sanitary City*; and Joel A. Tarr, *The Search for the Ultimate Sink: Urban Pollution in Historical Perspective* (Akron, Ohio: University of Akron Press, 1996).

12. Ted Steinberg, *Down to Earth: Nature's Role in American History* (New York: Ox-

ford University Press, 2002), 157–74; Tarr, *Search for the Ultimate Sink*, 292–98; and George Rogers Taylor, "The Beginnings of Mass Transportation in Urban America: Part II," *Smithsonian Journal of History* 1 (Autumn 1966): 31–54.

13. John Duffy, *The Sanitarians: A History of American Public Health* (Chicago: University of Illinois Press, 1992), 35–125.

14. For a discussion of how the courts dealt with nuisance issues during this period, see Christine Meisner Rosen, "'Knowing' Industrial Pollution: Nuisance Law and the Power of Tradition in a Time of Rapid Economic Change, 1840–1864," *Environmental History* 8 (October 2003): 565–97.

15. David Stradling, *Smokestacks and Progressives: Environmentalists, Engineers, and Air Quality in America, 1881–1951* (Baltimore, Md.: Johns Hopkins University Press, 1999).

16. Nancy S. Seasholes, *Gaining Ground: A History of Landmaking in Boston* (Cambridge: MIT Press, 2003); William Newman and Wilfred E. Holton, *Boston's Back Bay: The Story of America's Greatest Nineteenth-century Landfill Project* (Boston: Northeastern University Press, 2006); and Ann L. Buttenwieser, *Manhattan Water-Bound: Planning and Developing Manhattan's Waterfront from the Seventeenth Century to the Present* (New York: New York University Press, 1987).

17. David Schuyler, *The Redefinition of City Form in Nineteenth-century America* (Baltimore, Md.: Johns Hopkins University Press, 1986); and Roy Rosenzweig and Elizabeth Blackmar, *The Park and the People: A History of Central Park* (Ithaca, N.Y.: Cornell University Press, 1992).

18. Henry C. Binford, *The First Suburbs: Residential Communities on the Boston Periphery, 1815–1860* (Chicago: University of Chicago Press, 1985); and Kenneth T. Jackson, *The Crabgrass Frontier: The Suburbanization of the United States* (New York: Oxford, 1985), 3–102.

19. Melosi, *Sanitary City*, 90–99; and Tarr, *Search for the Ultimate Sink*, 180–90. Some of these urban streams are being exposed to the air today. See Richard Pinkham, *Daylighting: New Life for Buried Streams* (Old Snowmass, Colo.: Rocky Mountain Institute, 2000).

20. Tarr, *Search for the Ultimate Sink*, 111–218; and Melosi, *Sanitary City*, 90–116, 149–74.

21. Joel A. Tarr and Terry F. Yosie, "Critical Decisions in Pittsburgh Water and Wastewater Treatment," in *Devastation and Renewal: The Environmental History of Pittsburgh and Its Region*, edited by Joel A. Tarr (Pittsburgh: University of Pittsburgh Press, 2003), 76.

22. Arn Keeling, "Urban Wastes Sinks as a Natural Resource: The Case of the Fraser River," *Urban History Review* 34 (2005): 58–70; Tarr, *Search for the Ultimate Sink*, 354–73; and Richard White, *The Organic Machine: The Remaking of the Columbia River* (New York: Hill & Wang, 1995).

23. Tarr and Yosie, "Critical Decisions in Pittsburgh Water and Wastewater Treatment," 74–78; John O. Anfinson, *The River We Have Wrought: A History of the Upper*

Mississippi (Minneapolis: University of Minnesota Press, 2003), 149–51, 266–69; and John T. Cumbler, *Reasonable Use: The People, the Environment, and the State, New England, 1790–1930* (New York: Oxford University Press, 2001), 49–62, 131–60.

24. S. K. Schultz, *Constructing Urban Culture: American Cities and City Planning, 1800–1920* (Philadelphia: Temple University Press, 1989), 153–206; Neil Larry Shumsky, ed., *The Physical City: Public Space and Infrastructure* (New York: Garland, 1996); Raymond A. Mohl, *The New City: Urban America in the Industrial Age, 1860–1920* (Arlington Heights, Ill.: Harland Davidson, 1985), 26–42; and Robert M. Fogelson, *Downtown: Its Rise and Fall, 1880–1950* (New Haven, Conn.: Yale University Press, 2001).

25. Jackson, *Crabgrass Frontier*, 103–56; Dolores Hayden, *Building Suburbia: Green Fields and Urban Growth* (New York: Vintage Books, 2003), 71–127; and Tarr, *Search for the Ultimate Sink*, 303–22.

26. Robert Lewis, ed., *Manufacturing Suburbs: Building Work and Home on the Metropolitan Fringe* (Philadelphia: Temple University Press, 2004); Richard Harris, *Unplanned Suburbs: Toronto's American Tragedy, 1900–1950* (Baltimore, Md.: Johns Hopkins University Press, 1996); and Tarr, *Search for the Ultimate Sink*, 360–73.

27. Maureen A. Flanagan, *America Reformed: Progressives and Progressivisms, 1890s–1920s* (New York: Oxford University Press, 2006); and Martin V. Melosi, ed., *Pollution and Reform in American Cities, 1870–1930* (Austin: University of Texas Press, 1980).

28. Melosi, *Sanitary City*, 205–338; and Tarr, *Search for the Ultimate Sink*, 111–218.

29. Martin V. Melosi, *Garbage in the Cities: Refuse, Reform, and the Environment*, revised edition (Pittsburgh: University of Pittsburgh Press, 2005), 66–189.

30. Jackson, *Crabgrass Frontier*, 172–245; and Hayden, *Building Suburbia*, 97–153.

31. Tarr, *Search for the Ultimate Sink*, 323–34; and Clay McShane, *Down the Asphalt Path: The Automobile and the American City* (New York: Columbia University Press, 1994), 122.

32. Adam Rome, *The Bulldozer in the Countryside: Suburban Sprawl and the Rise of American Environmentalism* (New York: Cambridge University Press, 2001), 15–152.

33. Stradling, *Smokestacks and Progressives*, 85–191.

34. Samuel P. Hays, *Beauty, Health, and Permanence: Environmental Politics in the United States, 1955–1985* (New York: Cambridge University Press, 1987), 351–56; and Tarr, *Search for the Ultimate Sink*, 18–20.

35. Richard Harris and Michael E. Mercier, "How Healthy Were the Suburbs," *Journal of Urban History* 31 (September 2005): 767–98.

36. Hays, *Beauty, Health, and Permanence*, 70.

37. Richard N. L. Andrews, *Managing the Environment, Managing Ourselves: A History of American Environmental Policy* (New Haven, Conn.: Yale University Press, 1999), 227–54.

38. Sarah S. Elkind, "Los Angeles's Nature: Urban Environmental Politics in the Twen-

tieth Century," in *City, Country, Empire: Landscapes in Environmental History*, edited by Jeffry M. Diefendorf and Kurk Dorsey (Pittsburgh: University of Pittsburgh Press, 2005), 38–51; and Scott Hamilton Dewey, *Don't Breathe the Air: Air Pollution and U.S. Environmental Politics, 1945–1970* (College Station: Texas A&M Press, 2000), 37–112.

    39. Andrews, *Managing the Environment, Managing Ourselves*, 232–54.

    40. Hays, *Beauty, Health, and Permanence*, 76–83.

    41. Melosi, *Garbage in the Cities*, 190–226.

    42. Craig E. Colten and Peter N. Skinner, *The Road to Love Canal: Managing Industrial Waste before EPA* (Austin: University of Texas Press, 1996); and Niall Kirkwood, ed., *Manufactured Sites: Rethinking the Post-industrial Landscape* (New York: Spoon Press, 2001). For a provocative and skeptical study of how pollution issues were dealt with at Love Canal, see Allan Mazur, *A Hazardous Inquiry: The Rashomon Effect at Love Canal* (Cambridge: Harvard University Press, 1998).

    43. Andrews, *Managing the Environment, Managing Ourselves*, 255–318.

    44. The best source for automobile and transportation statistics are the annual reports of the U.S. Bureau of Transportation, available online from the Research and Innovative Technology Administration of the U.S. Department of Transportation, at http://www.bts.gov/publications/national_transportation_statistics/.

    45. Mark Rose, *Interstate Express Highway Politics, 1939–1989*, revised edition (Knoxville: University of Tennessee Press, 1990), 85–94.

    46. Hayden, *Building Suburbia*, 128–53; and Jackson, *Crabgrass Frontier*, 246–71.

    47. Hayden, *Building Suburbia*, 181–97; Robert Lang, *Edgeless Cities: Exploring the Elusive Metropolis* (Washington, D.C.: Brookings Institution Press, 2003); and Robert E. Lang and Jennifer LeFurgy, *Boomburbs: The Rise of America's Accidental Cities* (Washington, D.C.: Brookings Institution Press, 2007).

    48. Rome, *Bulldozer in the Countryside*; Elizabeth A. Johnson and Michael W. Klemens, eds., *Nature in Fragments: The Legacy of Sprawl* (New York: Columbia University Press, 2005); and F. Kaid Benfield, Matthew D. Raimi, and Donald D. T. Chen, *Once There Were Greenfields: How Urban Sprawl Is Undermining America's Environment, Economy, and Social Fabric* (Washington, D.C.: National Resources Defense Council, 1999).

    49. Owen D. Gutfreund, *Twentieth-century Sprawl: Highways and the Reshaping of the American Landscape* (New York: Oxford University Press, 2004). For an argument that "sprawl" reflects a historical trend that has more benefits than costs, see Robert Bruegmann, *Sprawl: A Compact History* (Chicago: University of Chicago Press, 2005).

    50. Carl Abbott, *The Metropolitan Frontier: Cities in the Modern American West* (Tucson: University of Arizona Press, 1995); Robert B. Fairbanks and Kenneth Underwood, eds., *Essays on Sunbelt Cities and Recent Urban America* (College Station: Texas A&M Press, 1990); Char Miller, ed., *On the Border: An Environmental History of San Antonio* (Pittsburgh: University of Pittsburgh Press, 2001); Michael F. Logan, *Desert Cities: The Environmental History of Phoenix and Tucson* (Pittsburgh: University of Pittsburgh Press,

2006); and William Deverell and Greg Hise, eds., *Land of Sunshine: An Environmental History of Metropolitan Los Angeles* (Pittsburgh: University of Pittsburgh Press, 2005).

51. See Tarr, *Devastation and Renewal*, for references.

52. There is a huge amount of literature on environmental justice. For bibliographical guidance, see Carolyn Merchant, *The Columbia Guide to American Environmental History* (New York: Columbia University Press, 2002); and Robin Lanette Turner and Diana Pei Wu, "Environmental Justice and Environmental Racism: An Annotated Bibliography and General Overview, Focusing on U.S. Literature, 1996–2002," Berkeley Workshop on Environmental Politics, Institute for International Studies, University of California, Berkeley, 134 pages; Bibliographies B 02–7, available online at http://globetrotter.berkeley.edu/EnvirPol/Bib/B07-TurnerWu.pdf.

53. The literature on sustainable cities is extensive. A useful guide is Kent E. Portney, *Taking Sustainable Cities Seriously: Economic Development, the Environment, and Quality of Life in American Cities* (Cambridge: Massachusetts Institute of Technology Press, 2003).

54. The literature on cities and the history of "natural disasters," after lying dormant for sometime, has been growing rapidly in recent years. It can now be expected to grow even further as a result of the events of Katrina and New Orleans. Some recent books include Ted Steinberg, *Acts of God: The Unnatural History of Natural Disaster in America* (New York: Oxford University Press, 2006); Craig Colten, *An Unnatural Metropolis: Wresting New Orleans from Nature* (Baton Rouge: Louisiana State University Press, 2005); James Orsi, *Hazardous Metropolis: Flooding and Urban Ecology in Los Angles* (Los Angeles: University of California Press, 2004); Ari Kelman, *A River and Its City: The Nature of Landscape in New Orleans* (Los Angles: University of California Press, 2003), especially chapter 5; John M. Barry, *Rising Tide: The Great Mississippi Flood of 1927 and How It Changed America* (New York: Touchstone Books, 1997); and Philip L. Fradkin, *The Great Earthquake and Firestorms of 1906: How San Francisco Nearly Destroyed Itself* (Berkeley: University of California Press, 2005). Still provoking is Mike Davis, *Ecology of Fear: Los Angeles and the Imagination of Disaster* (New York: Metropolitan Press, 1998). For Europe, see Genevieve Massard-Guilbaud, Harold L. Platt, and Dieter Schott, eds., *Cities and Catastrophes* (New York: Peter Lang, 2002).

55. Lawrence J. Vale and Thomas J. Campanella, *The Resilient City: How Modern Cities Recover from Disaster* (New York: Oxford University Press, 2005), 345–47; see Vale and Campanella's appendix "Suggestions for Further Reading on Urban Disasters and Recovery," 357–61.

56. Ibid., 350–51.

Chapter 7. The "1950s Syndrome" and the Transition from a Slow-Going to a Rapid Loss of Global Sustainability

1. Roger Revelle and Hans Suess, "Carbon Exchange between Atmosphere and Ocean and the Question of an Increase of Atmospheric CO_2 during the Past Decades," *Tellus* 9 (1957): 18–27.

2. Spencer R. Weart, *The Discovery of Global Warming: New Histories of Science, Technology, and Medicine* (Cambridge, Mass.: Harvard University Press, 2003), 30.

3. D. M. Etheridge, L. P. Steele, R. L. Langenfelds, R. J. Francey, J.-M. Barnola, and V. I. Morgan, "Historical CO_2 Records from the Law Dome DE08, DE08-2, and DSS Ice Cores," in *Trends: A Compendium of Data on Global Change* (Oak Ridge, Tenn.: Carbon Dioxide Information Analysis Center, Oak Ridge National Laboratory, U.S. Department of Energy, 1998), available online at http://cdiac.ornl.gov/trends/co2/lawdome.html (accessed on September 10, 2009). For measured CO_2 concentration at Mauna Loa since 1958, see ftp://ftp.cmdl.noaa.gov/ccg/co2/trends/co2_annmean_mlo.txt (accessed on September 10, 2009). This data is available online at http://www.japanfs.org/en/pages/029243.html (accessed on August 17, 2009).

4. The average increase was 0.3 ppm from 1900 to 1957. Dividing the aggregated increase of 79 ppm from 1957 to 2009 by the average increase from 1900 to 1957 yields 263 years!

5. Revelle as discussed in Al Gore, *Earth in the Balance: Ecology and the Human Spirit* (Boston: Houghton Mifflin, 2000), 5.

6. Translated from Hermann Flohn, *Physikalische Blätter* 37, no. 7 (1981): 184–90, quoted in Peter Weingart, Anita Engels, and Petra Pansegrau, "Risks of Communication: Discourses on Climate Change in Science, Politics, and the Mass Media," *Public Understanding of Science* 9, no. 3 (2000): 261–83, 268.

7. Will Steffen, Paul Crutzen, and John McNeill, "The Anthropocene: Are Humans Now Overwhelming the Great Forces of Nature?" *Ambio: A Journal of the Human Environment* 36, no. 8 (2007): 614–21, 616; Karl Polanyi, *The Great Transformation: The Political and Economic Origins of Our Time*, third edition, foreword by Joseph E. Stieglitz (Boston: Beacon Press, 2001).

8. Paul J. Crutzen, "Geology of Mankind: The Anthropocene," *Nature* 415 (2002): 23.

9. Steffen, Crutzen, and McNeill, "The Anthropocene."

10. Donella H. Meadows, Dennis L. Meadows, and Jørgen Randers, *Beyond the Limits: Confronting Global Collapse: Envisioning a Sustainable Future* (Post Mills, Vt.: Chelsea Green Publishing Co., 1992). Christian Pfister, "Das 1950er Syndrom: Der Energieverbrauch unserer Zivilisation in historischer Perspektive," *Natur und Mensch* 34, no. (1992): 1–4.

11. This collaboration took place at an interdisciplinary conference in Bern and was published as Christian Pfister, ed., *Das 1950er Syndrom: Der Weg in die Konsumgesellschaft*, second edition (Bern: Haupt Verlag, 1996), 179–200. Axel D. Neu, "Die Entfaltung der internationalen Erdölwirtschaft seit 1950," in *Das 1950er Syndrom*, edited by Pfister, 179–200. Christoph Maria Merki, "Der Treibstoffzoll aus historischer Sicht," in *Das 1950er Syndrom*, edited by Pfister, 311–32. Hans Flückiger, "Ramplanung im Spannungsfeld zwischen Trend und Steuerung," in *Das 1950er Syndrom*, edited by Pfister, 333–50. Gerhard Leutert, "Luftverschmutzung als Beispiel für Rückwirkungen auf die Umwelt," in *Das 1950er Syndrom*,

edited by Pfister, 351–76. Andreas Dieckmann, "Sozialer Wandel, Umweltbewusstsein, und Umweltverhalten," in *Das 1950er Syndrom*, edited by Pfister, 251–64. Marco Hüttenmoser, "Auswirkungen des '1950er Syndroms' auf den Alltag der Kinder," in *Das 1950er Syndrom*, edited by Pfister, 265–85. Klaus Armingeon, "Energiepolitik in Europa: Hindernisse umweltpolitischer Reformen," in *Das 1950er Syndrom*, edited by Pfister, 377–90. Wolf Linder, "Die fünfziger Jahre: Die Verarbeitung ökonomischer Modernisierung durch die politischen Institutionen der Schweiz," in *Das 1950er Syndrom*, edited by Pfister, 295–310.

12. "Der langfristige Rückgang der Relativpreise der fossilen Energieträger seit den späten 1950er Jahren ist die gewichtigste Ursache für den verschwenderischen Umgang mit Rohstoffen und Energie und die daraus erwachsenden übermässigen Belastungen der Umwelt" (Christian Pfister, Ruth Kaufmann-Hayoz, Beatrice Lanzrein, Gunter Stephan, Ewald C. Weibel, and Peter Gehr, "'Das 1950er Syndrom': Zusammenfassung und Synthese," in *Das 1950er Syndrom*, edited by Pfister, 21–47, 32).

13. Christian Pfister, "The 'Syndrome of the 1950s' in Switzerland: Cheap Energy, Mass Consumption, and the Environment," in *Getting and Spending: European and American Consumer Societies in the Twentieth Century*, edited by Susan Strasser, Charles McGovern, and Matthias Judt (New York: Cambridge University Press, 1998), 359–78.

14. Christian Pfister, "Energiepreis und Umweltbelastung: Zum Stand der Diskussion über das '1950er Syndrom,'" in *Umweltgeschichte: Themen und Perspektiven*, edited by Wolfram Siemann (Munich: Beck Verlag, 2003), 61–86.

15. Neither is it quoted in the corresponding more elaborate article by John McNeill, "Social, Economic, and Political Forces in Environmental Change, Dedal Scale (1900–2000)," in *Sustainability or Collapse? An Integrated History and Future of People on Earth*, edited by Robert Costanza, Lisa J. Graumlich, and Will Steffen (Cambridge, Mass.: MIT Press, 2007), 301–30.

16. The syndrome metaphor was taken up by Hans-Jürgen Schellnhuber and coworkers to designate other problems of nonsustainable development elsewhere, such as the "Dust Bowl Syndrome" (named after the 1935 Dust Bowl in the United States; see Hans-Jürgen Schellnhuber and others, "Syndromes of Global Change," *Gaia* 6, no. 1 [1997]: 19–34) or the "Favela Syndrome" in countries of the South (see Gerhard Petschel-Held and others, "Syndromes of Global Change: A Qualitative Modelling Approach to Assist Global Environmental Management," *Environmental Modelling and Assessment* [1998]: 1–21).

17. Christoph Maria Merki, "Epochenschwellen in der Wirtschafts-, Sozial- und Umweltgeschichte: Plädoyer für eine zeitliche Erweiterung von Pfisters '1950er Syndrom,'" in *Nachhaltige Geschichte: Festschrift für Christian Pfister*, edited by André Kirchhofer, Daniel Krämer, Christoph Maria Merki, Guido Poliwoda, Martin Stuber, and Stephanie Summermatter (Zurich: Chronos, 2009), 347–58, 355.

18. Jens Ivo Engels, "Umweltgeschichte als Zeitgeschichte," *Politik und Zeitgeschichte* 13 (2006): 32–38, quoted in Merki, "Epochenschwellen in der Wirtschafts-, Sozial- und Umweltgeschichte," 355.

19. Arne Andersen, *Der Traum vom guten Leben: Alltags- und Konsumgeschichte vom Wirtschaftswunder bis heute* (Frankfurt: Campus Verlag, 1997). Arne Andersen, "Das 50er-Jahre-Syndrom: Umweltfragen in der Demokratisierung des Technikkonsums," *Technikgeschichte* 65, no. (1998): 329–44.

20. Patrick Kupper, "Die '1970er Diagnose': Grundsätzliche Überlegungen zu einem Wendepunkt der Umweltgeschichte," *Archiv für Sozialgeschichte* 43 (2003): 325–48.

21. Axel Schildt, *Neue Politische Literatur* (1996), available online at http://www.ifs.tu-darmstadt.de/index.php?id=1450.

22. Hansjörg Siegenthaler, "Zur These des '1950er Syndroms': Die wirtschaftliche Entwicklung der Schweiz nach 1945 und die Bewegung relativer Energiepreise," in *Das 1950er Syndrom*, edited by Pfister, 97–105, 98. In doing so, he referred to my wording that cheap energy had "favored" the emergence of the "consumer society." See Christian Pfister, "Das '1950er Syndrom': Die umweltgeschichtliche Epochenschwelle zwischen Industriegesellschaft und Konsumgesellschaft," in *Das 1950er Syndrom*, edited by Pfister, 51–95, 77.

23. In his review of Pfister, *Das 1950er Syndrom*, Werner Abelshauser at least admits that this volume was an important step out of the ivory tower into the political arena (Abelshauser, *Archiv für Sozialgeschichte* 38 [1998]: 823–24, 823).

24. David G. Ockwell, "Energy and Economic Growth: Grounding Our Understanding in Physical Reality," *Energy Policy* 36 (2008): 4000–4400.

25. Peter Söderbaum, *Ecological Economics: A Political Economics Approach to Environment and Development* (London: Earthscan, 2001), 9–10 and 19. Significantly, Söderbaum does not mention history among the social sciences working within ecological economics, probably because historians thus far did not cooperate at all. This tergiversation may be explained with a deep-seated restive attitude of mainstream environmental history against economic arguments, just as Franz Mauelshagen diagnosed a tendency of opinion leaders in environmental history to avoid dealing with the approach of historical climatology. Franz Mauelshagen, "Keine Geschichte ohne Menschen: Die Erneuerung der historischen Klimawirkungsforschung aus der Klimakatastrophe," in *Nachhaltige Geschichte: Festschrift für Christian Pfister*, edited by André Kirchhofer, Daniel Krämer, Christoph Maria Merki, Guido Poliwoda, Martin Stuber, and Stephanie Summermatter (Zurich: Chronos, 2009), 169–93, 184.

26. Ockwell, "Energy and Economic Growth," 4001.

27. Nicholas Georgescu-Roegen, *The Entropy Law and the Economic Process* (Cambridge: Harvard University Press, 1971); and Juan Martinez-Alier and Klaus Schlüpmann, *Ecological Economics: Energy, Environment, and Society* (Oxford: Basil Blackwell, 1987). See also Ockwell, "Energy and Economic Growth," 4600–604.

28. Hans Christoph Binswanger and Elmar Ledergerber, "Bremsung des Energiezuwachses als Mittel der Wachstumskontrolle," in *Wirtschaftspolitik in der Umweltkrise*, edited by Jörg Wolff (Stuttgart: DVA, 1974), 104, quoted after Reiner Kümmel and Thomas Bruckner, "Energie, Entropie: Ökonomie, Ökologie," in *Das 1950er Syndrom*, edited by

Pfister, 129–148, 131 (see also Gunter Stephan, "'Das 1950er Syndrom' und Handlungsspielräume: Eine wirtschaftswissenschaftliche Betrachtung," in *Das 1950er Syndrom*, edited by Pfister).

29. Kümmel and Bruckner, "Energie, Entropie."

30. Rolf Peter Sieferle, Fridolin Krausmann, Heinz Schandl, and Verena Winiwarter, *Das Ende der Fläche: Zum gesellschaftlichen Stoffwechsel der Industrialisierung* (Cologne: Böhlau Verlag, 2006), 329.

31. Joachim Radkau, *Nature and Power: A Global History of the Environment* (Cambridge: Cambridge University Press, 2008), 252.

32. Pfister, "Die umweltgeschichtliche Epochenschwelle," 59. Merki, "Epochenschwellen in der Wirtschafts-, Sozial- und Umweltgeschichte," 355–58.

33. Siegenthaler, "Zur These des '1950er Syndroms,'" 100–102; and Radkau, *Nature and Power*, 285.

34. Frank Uekoetter, "Umweltbewegung zwischen dem Ende der nationalsozialistischen Herrschaft und der 'ökologischen Wende': Ein Literaturbericht," *Historical Social Research* 28 (2003): 270–89.

35. Steffen, Crutzen, and McNeill, "The Anthropocene."

36. Angus Maddison, *Contours of the World Economy, 1–2030 AD* (Oxford: Oxford University Press 2007), 71.

37. Ibid., 72.

38. Nicholas F. R. Crafts, "The Great Boom: 1950–73," in *Western Europe, Economic, and Social Change since 1945*, edited by Max-Stephan Schulze (London: Longman, 1999), 42–62, 42.

39. Hartmut Kaelble, *Der Boom, 1948–1973: Gesellschaftliche und wirtschaftliche Folgen in der Bundesrepublik Deutschland und in Europa* (Opladen: Westdeutscher Verlag, 1992), 12f.

40. Jean Fourastié, *Les trente glorieuses ou la révolution invisible* (Paris: Fayard, 1979). Crafts, "Great Boom," 42.

41. Burkart Lutz, *Der kurze Traum immerwährender Prosperität: Eine Neuinterpretation der industriell-kapitalistischen Entwicklung im Europa des 20. Jahrhunderts* (Frankfurt: Campus Verlag, 1989).

42. Maddison, *Contours of the World Economy*, 72

43. These population figures are from http://en.wikipedia.org/wiki/World_population#Population_figures (accessed on September 19, 2009).

44. Fridolin Krausmann, Simone Gingrich, Nina Eisenmenger, Karl-Heinz Erb, Helmut Haberl, and Marina Fischer-Kowalski, "Growth in Global Materials Use, GDP and Population during the Twentieth Century," *Ecological Economics* 68 (2009): 2696–705.

45. For many poor countries, data on GDP, sometimes even on population, are guesstimates, not considering the problems in converting the data from the original currencies to dollars and account for purchasing power parity (PPP).

46. Hans Bräker, "Die Tagungen des Rates für gegenseitige Wirtschaftshilfe und ihre

Ergebnisse (Teil 1)," *Osteuropa Wirtschaft* 7, no. 1 (1962): 58, as quoted in Benjamin Spycher, "Kampf gegen jede Verschwendung? Die Energieversorgung im Rat für gegenseitige Wirtschaftshilfe 1955 bis 1985 und die Auswirkungen auf die DDR nach 1975 am Beispiel des Erdöls" (M.A. thesis, Institute of History, University of Bern, 2008).

47. Crafts, "Great Boom," 61.

48. In late 1958 current-account currency convertibility was restored by all, except three, of the OEEC nations (Imanuel Wexler, *The Marshall Plan Revisited: The European Recovery Program in Economic Perspective* [Westport, Conn.: Greenwood Press, 1983], 254). Crafts, "Great Boom," 45.

49. On the Marshall Plan, see Wexler, *Marshall Plan Revisited*. On various forms of welfare, see Paul Johnson, "Welfare States," in *Western Europe: Economic and Social Change since 1945*, edited by Max-Stephan Schulze (London: Longman, 1999), 122–39. On educational reform, see Martin McLean, "Education," in *Western Europe*, edited by Schulze, 191–211.

50. Lutz, *Der kurze Traum immerwährender Prosperität*, 191–93; and Michael J. Piore and Charles F. Sabel, *Das Ende der Massenproduktion: Studie über die Requalifizierung der Arbeit und die Rückkehr der Ökonomie in die Gesellschaft* (Berlin: Klaus Wagenbach, 1985), 90–94.

51. Gabriel N. Alexander, "Impartial Umpireships: The General Motors–UAW Experience," in *Arbitration and the Law-Proceedings of the Twelfth Annual Meeting* (Washington, D.C.: National Academy of Arbitrators, 1959), 108–59.

52. On the establishment of welfare systems, see Johnson, "Welfare States," 122–23. On access to higher education, see McLean, "Education."

53. Crafts, "Great Boom," 43.

54. Knut Borchardt, "Die wirtschaftliche Entwicklung der Bundesrepublik nach dem 'Wirtschaftswunder,'" in *Der Weg der Bundesrepublik: Von 1945 bis zur Gegenwart*, edited by Franz Schneider (Munich: C. H. Beck Verlag, 1985), 193–216, 209.

55. Mohssen Massarat, "Erschöpfbare Ressourcen zu Dumpingpreisen: Wohlstand auf Kosten des Südens und künftiger Generationen," in *Das 1950er Syndrom*, edited by Pfister, 149–78.

56. Daniel Yergin, *The Prize: The Epic Quest for Oil, Money, and Power* (New York: Free Press, 1992), 591.

57. On the energy embargo by OPEC, see ibid., 735–47.

58. On the term "oil crisis," see Diona Venn, *The Oil Crisis* (London: Longman, 2002). For Germany, see Borchardt, "Die wirtschaftliche Entwicklung der Bundesrepublik nach dem 'Wirtschaftswunder,'" 202; and Klaus Voy, Werner Polster, and Claus Thomasberger, *Marktwirtschaft und politische Regulierung: Beiträge zur Wirtschafts- und Gesellschaftsgeschichte der Bundesrepublik Deutschland (1949–1949)* (Marburg: Metropolis-Verlag, 1991). For Switzerland, see Heidi Schelbert, "Schweizerische Wirtschaftsentwicklung seit 1950," in *Das 1950er Syndrom*, edited by Pfister, 201–18, 202–5.

59. Borchardt, "Die wirtschaftliche Entwicklung der Bundesrepublik nach dem 'Wirtschaftswunder,'" 202.

60. Siegwart Lindenberg and Bruno S. Frey, "Alternatives, Frames, and Relative Prices: A Broader View of Rational Choice Theory," *Acta Sociologica* 36, no. 3 (1993): 191–205, 191.

61. Dealing with this issue in somewhat more detail should not be seen as legitimizing the neoclassical construct of the "economic man" that is now seen to have engendered most environmental problems, last but not least regarding global warming. Although ecological sustainability is of primary importance, Peter Söderbaum's argument is convincing: we should not throw out the baby with the bathwater by distancing ourselves from arguments of neoclassical economics, which still have their uses. Söderbaum, *Ecological Economics*, chapter 7.

62. Hansjörg Siegenthaler, *The Regulation School: A Critical Introduction* (New York: Columbia University Press, 1990).

63. See Lindenberg and Frey, "Alternatives, Frames, and Relative Prices," 191.

64. Hansjörg Siegenthaler, *Regelvertrauen, Prosperität, und Krisen: Die Ungleichmässigkeit wirtschaftlicher und sozialer Entwicklung als Ergebnis individuellen Handelns und sozialen Lernens* (Tübingen: J. C. B. Mohr [Paul Siebeck], 1993).

65. Possibly the best study on this issue is still Helmut J. Frank, *Crude Oil Prices in the Middle East: A Study in Oligopolistic Price Behavior* (New York: Praeger, 1966).

66. Yergin, *The Prize*, 393.

67. Matt Simmons, *Twilight in the Desert: The Coming Saudi Oil Shock and the World Economy* (Hoboken, N.J.: John Wiley, 2005), 151–52.

68. "Saudi Aramco History," available online at http://www.jobsataramco.com/Home/history.aspx (accessed on November 3, 2009).

69. Yergin, *The Prize*; Anthony Sampson, *The Seven Sisters* (Sevenoaks, England: Coronet, 1988).

70. Frank, *Crude Oil Prices in the Middle East*, 128.

71. John Blair, *The Control of Oil* (London: Macmillan, 1976), 47 and following pages, quoted in Peter von Allmen, "Energie und Europa—Kohle, Öl, und das 'friedliche' Atom" (Ph.D. dissertation, Bern, 2002), 158, available online at http://www.stub.unibe.ch/download/eldiss/02allmen_p.pdf (accessed on September 19, 2009).

72. Martin Chick, *Electricity and Energy Policy in Britain, France, and the United States since 1945* (Cheltenham: Edward Elgar, 2007), 8–9.

73. Sampson, *Seven Sisters*, as quoted in Jacoba van der Linden, *Dynamic International Oil Markets* (Dordrecht: Kluwer Academic Publishers, 1991), 147n72.

74. Edit Penrose, *The Large International Firm in Developing Countries: The International Petroleum Industry* (London: G. Allen and Unwin, 1968), 79.

75. Massarat, "Erschöpfbare Ressourcen zu Dumpingpreisen," 166–71.

76. Von Allmen, "Energie und Europa"; and Chick, *Electricity and Energy Policy*, 8–10.

77. Frank, *Crude Oil Prices in the Middle East*, 142. On the Hartley Report, see Organi-

zation for European Economic Co-operation, *Europe's Growing Needs of Energy: How Can They Be Met?* (Paris: OEEC, 1956).

78. Andersen, *Der Traum vom guten Leben*, 765.

79. Ian Seymour, *OPEC: Instrument of Change* (London: Macmillan Press, 1980).

80. Von Allmen, "Energie und Europa," 135–41.

81. Borchardt, "Die wirtschaftliche Entwicklung der Bundesrepublik nach dem 'Wirtschaftswunder,'" 209.

82. Von Allmen, "Energie und Europa," 135–41.

83. Werner Abelshauser, *Der Ruhrkohlebergbau seit 1945: Wiederaufbau, Krise, Anpassung* (Munich: C. H. Beck Verlag, 1992), 110.

84. Frank, *Crude Oil Prices in the Middle East*, 86.

85. For the United Kingdom, see John Sheail, *An Environmental History of Twentieth-century Britain* (Basingstoke: Palgrave Macmillan, 2002), 204–11.

86. Richard H. K. Vietor, *Contrived Competition: Regulation and Deregulation in America* (Cambridge, Mass.: Harvard University Press, 1994), 95, as quoted in Chick, *Electricity and Energy Policy*, 8.

87. Richard H. K. Vietor, *Energy Policy in America since 1945: A Study of Business-Government Relations* (Cambridge: Cambridge University Press, 1984), 347.

88. Neu, "Die Entfaltung der internationalen Erdölwirtschaft seit 1950," estimated from figure 29 on page 185. Prices are in US$ from 1991.

89. Georg Erdmann, *Energieökonomik: Theorie und Anwendungen* (Zurich: B. G. Teubner Verlag, 1992).

90. Peter Brimblecombe, *The Big Smoke: A History of Air Pollution in London since Medieval Times* (London: Methuen, 1987).

91. D. J. Stern and C. Cleveland, *Energy and Economic Growth*. Rensselaer Working Paper in Economics No. 0410 (Troy, N.Y.: Rensselaer Polytechnic Institute), as quoted in Ockwell, "Energy and Economic Growth," 4602.

92. Von Allmen, "Energie und Europa," 119–21.

93. Klaus Tenfelde, *Sozialgeschichte der Bergarbeiterschaft an der Ruhr im 19. Jahrhundert* (Bonn: Verlag Neue Gesellschaft, 1981), 292; and Abelshauser, *Der Ruhrkohlebergbau seit 1945*, 72.

94. Von Allmen, "Energie und Europa," 90–167.

95. Ibid., 125–26.

96. Abelshauser, *Der Ruhrkohlebergbau seit 1945*, 92.

97. John G. Clark, *The Political Economy of World Energy: A Twentieth Century Perspective* (New York: Harvester-Wheatsheaf, 1990), 25.

98. Chick, *Electricity and Energy Policy*, 9.

99. Abelshauser, *Der Ruhrkohlebergbau seit 1945*, 72; and Chick, *Electricity and Energy Policy*, 13.

100. Thomas G. Weyman-Jones, *Energy in Europe* (London: Methuen, 1986), 55.

101. Boris Rachkov, "Oil, Trade, and Politics," *International Affairs* 12 (1966): 14–20, 14, as quoted in Christian Scheidegger, "Ursachen für den Ölpreisfall ab 1957–58; Veränderungen im Erdölmarkt der Nachkriegszeit und deren Rezeption in der westeuropäischen Politik und Wirtschaft," (Bern, 2003).

102. Harry Fornari, *Bread upon the Waters: A History of the United States Grain Exports* (Nashville, Tenn.: Aurora Publishers, 1973).

103. Uwe Röndigs, *Globalisierung und europäische Integration: Der Strukturwandel des Energiesektors und die Politik der Montanunion, 1952–1962* (Baden-Baden: F. Steiner Verlag, 1996, 2000). Jens Hohensee, *Der erste Ölpreisschock, 1973–74* (Stuttgart, 1976). Venn, *Oil Crisis*.

104. Joszef M. Brabant, *Economic Integration in Eastern Europe* (New York: Routledge, 1989), 320.

105. Jennifer L. Considine and William A. Kerr, *The Russian Oil Economy* (Cheltenham: Edward Elgar, 2002), 63. Dumping prices paid by noncommunist countries still were substantially lower (ibid., 82). COMECON officials decided in 1958 to take average world market prices for reference adjusting export prices to member states in each case after a couple of years (Brabant, *Economic Integration in Eastern Europe*, 320). Bräker, as quoted in Spycher, "Kampf gegen jede Verschwendung?"

106. Considine and Kerr, *Russian Oil Economy*, 137; and Margaret Chadwick, David Long, and Machiko Nissanke, *Soviet Oil Exports: Trade Adjustments, Refining Constraints, and Market Behaviour* (Oxford: Oxford University Press, 1987), 36.

107. Spycher, "Kampf gegen jede Verschwendung?" 65–66.

108. Petra Kuskova, Simone Gingrich, and Fridolin Krausmann, "Long-term Changes in Social Metabolism and Land Use in Czechoslovakia, 1830–2000: An Energy Transition under Changing Political Regimes," *Ecological Economics* 68 (2008): 394–407.

109. Otto Kandler and John L. Innes, "Air Pollution and Forest Decline in Central Europe," *Environmental Pollution* 90 (1995): 171–80.

110. Chick, *Electricity and Energy Policy*, 7.

111. David Popp, "Induced Innovation and Energy Prices," *American Economic Review* 92, no. 1 (2002): 160–80, 170.

112. Richard G. L. Newell, Adam B. Jaffe, and Robert N. Stavins. *The Induced Innovation Hypothesis and Energy-Saving Technological Change*. NBER Working Paper 6437 (Cambridge, Mass.: National Bureau of Economic Research, 1998), as quoted in Robert Fouquet, *Heat, Power, and Light: Revolutions in Energy Services* (Cheltenham: Edward Elgar, 2008), 19.

113. Popp, "Induced Innovation and Energy Prices," 160.

114. Chris Cook and John Stevenson, *Longman Atlas of Modern British History: A Visual Guide to British Society and Politics, 1700–1970* (London: Longman, 1978), 38.

115. Fouquet, *Heat, Power, and Light*, 121.

116. Ken Butti and John Perlin, *A Golden Thread: 2500 Years of Solar Architecture and Technology* (New York: Van Nostrand Reinhold, 1980), 233, 238, and 247.

117. Ibid., 195.

118. With regard to Europe this term was coined by Victor Scardigli, *La consummation: Culture du quotidien* (Paris: Presses Universitaires de France, 1983), as quoted in Christof Haudenschild, *Konsum als Lebensweltstilisierung* (Bern: Peter Lang Verlag, 1989), 258.

119. Wolfgang König, *Geschichte der Konsumgesellschaft* (Stuttgart, 2000), 108.

120. See Gary Cross, *Time and Money: The Making of Consumer Culture* (London: Routledge, 1993), 1.

121. König, *Geschichte der Konsumgesellschaft*, 23.

122. Victoria De Grazia, "Changing Consumption Regimes in Europe, 1930–1970," in *Getting and Spending: European and American Consumer Societies in the Twentieth Century*, edited by Susan Strasser, Charles McGovern, and Matthias Judt (Cambridge: Cambridge University Press, 1998), 59–84.

123. Michael Wildt, "Consumption as Social Practice in West Germany," in *Getting and Spending*, edited by Strasser, McGovern, and Judt, 301–16, 313.

124. Ibid., 303.

125. Pfister, "Die umweltgeschichtliche Epochenschwelle," 72.

126. Wildt, "Consumption as Social Practice in West Germany," 307.

127. Sabine Haustein, *Vom Mangel zum Massenkonsum: Deutschland, Frankreich, und Grossbritannien im Vergleich, 1945–1970* (Frankfurt: Campus Verlag, 2007), 72–80.

128. Pfister, "Die umweltgeschichtliche Epochenschwelle," 65; and Haustein, *Vom Mangel zum Massenkonsum*, 81–115.

129. Pfister, "'Syndrome of the 1950s' in Switzerland," 363–64.

130. Lutz, *Der kurze Traum immerwährender Prosperität*; and Pfister, "'Syndrome of the 1950s' in Switzerland," 364.

131. Ursula Schroth-Pritzel, "Der Arbeitszeitaufwand im städtischen Haushalt," *Hauswirtschaft und Wissenschaft* 4 (1958): 7–22, as quoted in Wildt, "Consumption as Social Practice in West Germany," 307.

132. Pfister, "Die umweltgeschichtliche Epochenschwelle," 65.

133. Mathis Wackernagel, "Using 'Appropriated Carrying Capacity' as an Indicator: Measuring the Sustainability of a Community," Technical Report to the UBC Task Force on Healthy and Sustainable Communities, UBC School of Community and Regional Planning, Vancouver, 1991. William E. Rees, "Ecological Footprints and Appropriated Carrying Capacity: What Urban Economics Leaves Out," *Environment and Urbanization* 4 (1992): 121–30. Mathis Wackernagel, "Methodological Advancements in Footprint Analysis," *Ecological Economics* 68 (2009): 1925–27.

134. See "ecological footprint" at the Global Footprint Network at http://www.footprintnetwork.org/en/index.php/GFN/page/frequently_asked_questions/#gen2 (accessed on August 27, 2009).

135. Maddison, *Contours of the World Economy*, here and there.

136. Matthias Nast, *Die stummen Verkäufer: Lebensmittelverpackungen im Zeitalter der Konsumgesellschaft. Umwelthistorische Untersuchung über die Entwicklung der Warenverpackungen und den Wandel der Einkaufsgewohnheiten (1950er bis 1990er Jahre)* (Bern: Peter Lang Verlag, 1997).

137. Sieferle et al., *Das Ende der Fläche*.

138. Vaclav Smil, *Energy in World History* (Boulder, Colo.: Westview Press, 1994), 42–45.

139. Paul Bairoch, "Évolutions Économiques : Les Trois Révolutions Agricoles du Monde Développé. Rendements et Productivité de 1800 à 1985," *Annales ESC* 2 (1989): 317–53.

140. John McNeill, *Something New under the Sun: An Environmental History of the Twentieth Century* (London: W. W. Norton, 2000), 216.

141. Sieferle et al., *Das Ende der Fläche*.

142. McNeill, *Something New under the Sun*, 219–26.

143. Hermann Priebe, *Die subventionierte Naturzerstörung: Plädoyer für eine neue Agrarkultur* (Munich: Goldmann, 1990), 13pp., translation from the original German is mine.

144. Pfister, "'Syndrome of the 1950s' in Switzerland," 364.

145. Verena Winiwarter, "Landwirtschaft, Natur, und ländliche Gesellschaft im Umbruch: Eine umwelthistorische Perspektive zur Agrarmodernisierung," in *Agrarmodernisierung und ökologische Folgen: Westfalen vom 18. bis zum 20. Jahrhundert*, edited by Karl Ditt, Rita Gudermann, and Norwich Rüsse (Paderborn: Ferdinand Schöningh Verlag, 2001), 733–67, 733.

146. Fridolin Krausmann and Helmut Haberl, "The Process of Industrialization from the Perspective of Energetic Metabolism: Socioeconomics Energy Flows in Austria, 1830–1995," *Ecological Economics* 41, no. 2 (2002): 177–201.

147. Tyler G. Miller, *Living in the Environment* (Belmont, Calif.: Brooks/Cole, 2002); and McNeill, *Something New under the Sun*.

148. Eric Hobsbawm, *Das Zeitalter der Extreme* (Munich: Carl Hanser Verlag, 1995), 365.

149. McNeill, *Something New under the Sun*, 212–26.

150. Giovanni Federico, *Feeding the World: An Economic History of Agriculture, 1800–2000* (Princeton, N.J.: Princeton University Press 2005), 24.

151. Christian Pfister, "Die Umwälzung der Landwirtschaft im Tieferen [Schweizer] Mittelland seit den 1950er Jahren und der Untergang des Bauerntums," in *Die Bauernhäuser des Kantons Bern*, vol. 3, edited by Heinrich Christoph Affolter (Bern: In press).

152. Michael Tracy, *Government and Agriculture in Western Europe, 1880–1988* (New York: New York University Press, 1989), 224–25, 243–76; and Jörn Sieglerschmidt, "Die Industrialisierung der landwirtschaftlichen Produktion seit 1950 in *Der Aufbruch ins Schlaraffenland*," *Environmental History Newsletter*, special issue 2 (1995): 181–203.

153. Vaclav Smil, *Enriching the Earth: Fritz Haber, Carl Bosch, and the Transformation of World Food Production* (Cambridge, Mass.: MIT Press, 2004), 203 Also see http://www.agmrc.org/renewable_energy/ethanol/energy_agriculture__wheres_the_nitrogen.cfm (accessed on August 29, 2009).

154. See, for example, Federico, *Feeding the World*.

155. Haustein, *Vom Mangel zum Massenkonsum*, 55–80, 116–32.

156. Antoinette Mannion, *Global Environmental Change* (Essex: Longman, 1997). For Germany, see Frank Uekoetter, *Umweltgeschichte im 19. und 20. Jahrhundert* (Munich: R. Oldenbourg, 2007), 115–16.

157. Werner Polster, "Wandlungen der Lebensweise im Spiegel der Konsumentwicklung—Vom Dienstleistungskonsum zum demokratischen Warenkonsum," in *Gesellschaftliche Transformationsprozesse und materielle Lebensweise: Beiträge zur Wirtschafts- und Gesellschaftsgeschichte der Bundesrepublik Deutschland (1949–1989)*, edited by K. Voy, W. Polster, and C. Thomasberger (Marburg: Metropolis-Verlag, 1991), 215–91.

158. Peter Borscheid, "Auto und Massenmobilität," in *Die Einflüsse der Motorisierung auf das Verkehrswesen von 1886 bis 1986, Zeitschrift für Unternehmensgeschichte, Beiheft 52*, edited by Hans Pohl (Wiesbaden: Franz Steiner Verlag, 1988), 117–41, 122.

159. Philip Bagwell, *The Transport Revolution, 1770–1985* (London: Routledge, 1988), 353.

160. Christoph Kopper, *Handel und Verkehr im 20. Jahrhundert* (Munich: Oldenbourg Verlag, 2002), 58. Joachim Radkau has observed that "the increase in the simple number of cars is an amorphic process which is not adequate for assessing its environmental impact." Of course, a classification of cars according to mileage would be more suitable as a statistic, but this data is not available. Joachim Radkau, *Technik in Deutschland: Vom 18. Jahrhundert bis zur Gegenwart* (Frankfurt: Suhrkamp Verlag, 1989).

161. Christoph Maria Merki, *Der holprige Siegeszug des Automobils, 1895–1930* (Vienna: Böhlau Verlag, 2002), 21–23.

162. John Armstrong, "Transport and Communications," in *Western Europe: Economic and Social Change since 1945*, edited by Max-Stephan Schulze (London: Longman, 1999), 212–33, 218.

163. Gijs Mom, "Roads without Rails: European Highway-Network Building and Desire for Long-Range Motorized Mobility," *Technology and Culture* 46 (2005): 745–72, 748.

164. Ibid., 771.

165. Ibid., 760.

166. Ibid., 748.

167. Ibid., 771.

168. Armstrong, "Transport and Communications," 231.

169. The right to vote was not granted to women before 1971. Merki, "Der Treibstoffzoll aus historischer Sicht."

170. Mom, "Roads without Rails," 765.

171. For Switzerland, see Merki, "Der Treibstoffzoll aus historischer Sicht."

172. Linder, "Die fünfziger Jahre." For Germany, see Kopper, *Handel und Verkehr im 20. Jahrhundert*, 62.

173. Pfister, "Die umweltgeschichtliche Epochenschwelle."

174. *Die Weltwoche*, November 18, 1973.

175. Pfister, "Die umweltgeschichtliche Epochenschwelle."

176. Armstrong, "Transport and Communications," 212.

177. Ibid., 213.

178. Ibid., 213, 220, data are added up from table 13.4.

179. Hans-Liudger Dienel and Peter Lyth, *Flying the Flag: European Commercial Air Transport since 1945* (Basingstoke: Macmillan Press, 1998), 3.

180. This material comes from "Boeing 707 Turbojet Airliner," available online at http://www.historylink.org/index.cfm?DisplayPage=output.cfm&File_Id=3890 (accessed on September 22, 2009).

181. Armstrong, "Transport and Communications," 218.

182. Yergin, *The Prize*, 884–91.

183. Armstrong, "Transport and Communications," 230.

184. Dipesh Chakrabarty, "The Climate of History: Four Theses," *Critical Inquiry* 35 (2009): 197–222.

185. Nicholas Stern, *The Economics of Climate Change: The Stern Review* (Cambridge: Cambridge University Press, 2007).

186. "Saudi Aramco History," data available online at http://www.jobsataramco.com/Home/history.aspx (accessed on November 3, 2009).

187. M. King Hubbert, "Energy from Fossil Fuels," *Science* 109 (1949): 103–9.

188. "M. King Hubbert," available online at http://www.hubbertpeak.com/Hubbert/ (accessed on December 2, 2009).

Chapter 8. Modern Environmentalism

1. See, for example, James Lovelock, *Gaia: A New Look at Live on Earth* (Oxford: Oxford University Press, 1979).

2. Franz-Josef Brüggemeier, "Umweltgeschichte: Erfahrungen, Ergebnisse, Erwartungen," *Archiv für Sozialgeschichte* 43 (2003): 1–18. For an overview of environmental history, see John R. McNeill, "Observations on the Nature and Culture of Environmental History," *History and Theory* (theme issue) 42 (2003): 5–43.

3. Ulrich Beck, *Risikogesellschaft: Auf dem Weg in eine andere Moderne* (Frankfurt: Suhrkamp, 1986); Ulrich Beck, Anthony Giddens, and Scott Lash, *Reflexive Modernization: Politics, Tradition, and Aesthetics in the Modern Social Order* (Cambridge: Polity Press, 1994); and Ulrich Beck, ed., *Theorie und Empirie reflexiver Modernisierung* (Baden-Baden: Namos, 2005).

4. See Franz-Josef Brüggemeier and Jens Ivo Engels, "Den Kinderschuhen entwach-

sen: Einleitende Worte zur Umweltgeschichte der zweiten Hälfte des 20. Jahrhunderts," in *Natur- und Umweltschutz in Deutschland nach 1945: Konzepte, Konflikte, Kompetenzen*, edited by Franz-Josef Brüggemeier and Jens Ivo Engels (Frankfurt: Campus, 2005), 10–19.

5. Kai F. Hünemörder, "Kassandra im modernen Gewand: Die umweltapokalyptischen Mahnrufe der frühen 1970er Jahre," in *Wird Kassandra heiser? Die Geschichte falscher Ökoalarme*, edited by Frank Uekötter and Jens Hohensee (Stuttgart: Franz Steiner, 2004), 78–97.

6. Rainer Geißler, *Die Sozialstruktur Deutschlands*, third edition (Bonn: Bundeszentrale für politische Bildung, 2002), 60; and Richard Münch, "Die 'Zweite Moderne': Realität oder Fiktion? Kritische Fragen an die Theorie der 'reflexiven' Modernisierung," *Kölner Zeitschrift für Soziologie und Sozialpsychologie* 54 (2002): 417–43.

7. Björn Lomborg, *The Skeptical Environmentalist: Measuring the Real State of the World* (Cambridge: Cambridge University Press, 2001).

8. See Christian Pfister's contribution in this volume.

9. Christian Pfister, ed., *Das 1950er Syndrom: Der Weg in die Konsumgesellschaft* (Bern: Paul Haupt, 1995).

10. Umweltbundesamt, ed., *Daten zur Umwelt, 1986–87*, second edition (Berlin: Schmidt, 1986), 232–33.

11. Umweltbundesamt, ed., *Daten zur Umwelt 1992–93* (Berlin: Schmidt, 1994), 327. See also Franz-Josef Brüggemeier, *Tschernobyl, 26. April 1986. Die ökologische Herausforderung* (Munich: dtv, 1998).

12. See Rolf-Peter Sieferle, *The Subterranean Forest: Energy Systems and the Industrial Revolution* (Cambridge: White Horse Press, 2001); and J. R. McNeill, *Something New under the Sun: An Environmental History of the Twentieth-century World* (New York: Norton, 2000).

13. With respect to pollution control, see Franz-Josef Brüggemeier, *Das unendliche Meer der Lüfte: Luftverschmutzung, Industrialisierung, und Risikodebatten im 19. Jahrhundert* (Essen: Klartext, 1996); Frank Uekoetter, *Von der Rauchplage zur ökologischen Revolution: Eine Geschichte der Luftverschmutzung in Deutschland und den USA, 1880–1970* (Essen: Klartext, 2003); and Stephen Mosley, *The Chimney of the World: A History of Smoke Pollution in Victorian and Edwardian Manchester* (Cambridge: White Horse Press, 2001).

14. Dennis L. and Donnella H. Meadows, *The Limits to Growth: A Report for the Club of Rome's Project on the Predicament of Mankind* (London: Earth Island, 1972).

15. Patrick Kupper, "Die '1970er Diagnose': Grundsätzliche Überlegungen zu einem Wendepunkt der Umweltgeschichte," *Archiv für Sozialgeschichte* 43 (2003): 325–48.

16. See the study by Kai F. Hünemörder, *Die Frühgeschichte der globalen Umweltkrise und die Formierung der deutschen Umweltpolitik (1950–1973)* (Stuttgart: Steiner, 2004). On West German futurology and its role in politics, see Alexander Schmidt-Gernig, "Die gesellschaftliche Konstruktion der Zukunft: Westeuropäische Zukunftsforschung und

Gesellschaftsplanung zwischen 1950 und 1980," *WeltTrends: Zeitschrift für internationale Politik und vergleichende Studien* 18 (1998): 63–84.

17. Dieter Rucht, *Modernisierung und neue soziale Bewegungen: Deutschland, Frankreich, und USA im Vergleich* (Frankfurt: Campus, 1994).

18. Brüggemeier, *Das unendliche Meer*; and Uekoetter, *Von der Rauchplage*.

19. "Umwelt als Bumerang," *Frankfurter Allgemeine Zeitung*, August 5, 1970.

20. Holger Strohm, *Politische Ökologie: Arbeitsmaterialien und Lernmodelle für Unterricht und Aktion* (Reinbek: Rowohlt, 1979).

21. Anna Hallensleben, *Von der Grünen Liste zur Grünen Partei* (Göttingen: Muster-Schmidt, 1984), including original documents.

22. Björn-Ola Linnér, *The Return of Malthus: Environmentalism and Post-war Population-Resource Crises* (Cambridge: White Horse Press, 2003).

23. Kai Hünemörder, "Vom Expertennetzwerk zur Umweltpolitik: Frühe Umweltkonferenzen und die Ausweitung der öffentlichen Aufmerksamkeit für Umweltfragen in Europa (1959–1972)," *Archiv für Sozialgeschichte* 43 (2003): 275–96.

24. Anna-Katharina Wöbse, "Der Schutz der Natur im Völkerbund: Beginn einer Weltumweltpolitik," *Archiv für Sozialgeschichte* 43 (2003): 177–90.

25. Fairfield Osborn, *Our Plundered Planet* (Boston: Little, Brown and Co., 1948). Pierre Desrochers and Christine Hoffbauer, "The Post-war Intellectual Roots of the Population Bomb: Fairfield Osborn's 'Our Plundered Planet' and William Vogt's 'Road to Survival' in Retrospect," *Electronic Journal of Sustainable Development* 1 (2009): 37–61. On the doomsday literature of the 1940s and 1950s and their perception in West Germany, see Jens Ivo Engels, *Naturpolitik in der Bundesrepublik: Ideenwelt und politische Verhaltensstile in Naturschutz und Umweltbewegung, 1950–1980* (Paderborn: Schöningh, 2006), chapter 1.

26. Konrad Lorenz, *Die acht Todsünden der zivilisierten Menschheit* (Munich: Pieper, 1973); see Klaus Taschwer and Benedikt Föger, *Konrad Lorenz: Biographie* (Vienna: Zsolnay, 2003).

27. Engels, *Naturpolitik in der Bundesrepublik*, chapter 10.

28. Ibid.

29. Franz-Josef Brüggemeier, "Waldsterben: The Construction and Deconstruction of an Environmental Problem," in *Nature in German History*, edited by Christof Mauch (New York: Berghahn, 2004), 119–31; Frank Uekoetter and Kenneth Anders, "Viel Lärm ums stille Sterben: Die Debatte über das Waldsterben in Deutschland," in *Wird Kassandra heiser? Die Geschichte falscher Ökoalarme*, edited by Frank Uekoetter and Jens Hohensee (Stuttgart: Franz Steiner, 2004), 112–38.

30. Udo Kuckartz, "Umweltwissen, Umweltbewußtsein, Umweltverhalten: Der Stand der Umweltbewußtseinsforschung," in *Umweltbewußtsein und Massenmedien: Perspektiven ökologischer Kommunikation*, edited by Gerhard de Haan (Berlin: Akademie, 1995), 71–85.

31. Walter Hollstein, *Die Gegengesellschaft: Alternative Lebensformen*, second edition

(Bonn: Neue Gesellschaft, 1980); and Sabine von Dirke, *"All Power to the Imagination!" The West German Counterculture from the Student Movement to the Greens* (London: Nebraska University Press, 1997).

32. Matthias Heymann, "Luftverschmutzung, Atmosphärenforschung, Luftreinhaltung: Ein technisches Problem?" in *Natur- und Umweltschutz nach 1945: Konzepte, Konflikte, Kompetenzen*, edited by Franz-Josef Brüggemeier and Jens Ivo Engels (Frankfurt: Campus, 2005), 325–42.

33. Helmut Kohl, quoted in *Verhandlungen des Deutschen Bundestages: Stenographische Berichte* 189 (1997): 16,826.

34. Münch, "Die 'Zweite Moderne': Realität oder Fiktion?"

Chapter 9. The Knowledge Society

1. This interpretation is greatly indebted to the reading by Wolfgang Krohn, *Francis Bacon* (Munich: Beck, 1987).

2. Peter Burke, *A Social History of Knowledge* (Cambridge: Polity Press, 1997).

3. Carolyn Merchant, *The Death of Nature: Women, Ecology, and the Scientific Revolution* (San Francisco: Harper & Row, 1980), 1.

4. See Michael Dettelbach, "Alexander von Humboldt zwischen Aufklärung und Romantik," in *Alexander von Humboldt: Aufbruch in die Moderne*, edited by Ottmar Ette et al. (Berlin: Akademie-Verlag, 2001), 137–49.

5. Ellis Spedding and London Heath, eds., *The Works of Francis Bacon*, Volume 1 (1858, reprint; St. Clair Shores, Mich.: Scholarly Press, 1963), 157. Translation from Francis Bacon, *The New Organon and Related Writings*, edited, with an introduction, by Fulton H. Anderson (New York: Liberal Arts Press, 1960), 39.

6. Bennett, as quoted in National Archives of the United States Record Group (RG) 114, Entry 1021, Box 2, Folder "Illinois," The Mission of Soil Conservation, address by Dr. H. H. Bennett, chief of the Soil Conservation Service, during the National Convocation on The Church in Town and Country, at Elgin, Illinois, November 16, 1944, page 1.

7. Hans-Luidger Dienel, *Herrschaft über die Natur? Naturvorstellungen deutscher Ingenieure, 1871–1914* (Stuttgart: Verlag für Geschichte der Naturwissenschaften und der Technik, 1992).

8. See Allan M. Winkler, *Life under a Cloud: American Anxiety about the Atom* (Urbana: University of Illinois Press, 1999).

9. See Joachim Radkau, *Aufstieg und Krise der deutschen Atomwirtschaft, 1945–1975: Verdrängte Alternativen und der Ursprung der nuklearen Kontroverse* (Reinbek: Rowohlt, 1983).

10. Jochim Varchmin and Joachim Radkau, *Kraft, Energie, und Arbeit: Energie und Gesellschaft* (Reinbek bei Hamburg: Rowohlt, 1981), 93n.

11. Geheimes Staatsarchiv Preußischer Kulturbesitz, Rep. 120 BB II a 2 No. 28 Adh. 1, vol. 4, p. 194.

12. Ulrich Wengenroth, "Eiffelturm und Coladose: Über den Zusammenhang von Wirtschaftswachstum und Metallverbrauch," *Kultur & Technik* 19, no. 3 (1995): 14.

13. Quoted after Danny Trom, "Natur und nationale Identität: Der Streit um den Schutz der 'Natur' um die Jahrhundertwende in Deutschland und Frankreich," in *Nation und Emotion: Deutschland und Frankreich im Vergleich. 19. und 20. Jahrhundert*, edited by Etienne François, Hannes Siegrist, and Jakob Vogel (Göttingen: Vandenhoeck und Ruprecht, 1995), 147.

14. See Martin V. Melosi, *The Sanitary City: Urban Infrastructure in America from Colonial Times to the Present* (Baltimore, Md.: Johns Hopkins University Press, 2000).

15. See Timothy LeCain, "The Limits of 'Eco-Efficiency': Arsenic Pollution and the Cottrell Electrical Precipitator in the U.S. Copper Smelting Industry," *Environmental History* 5 (2000): 336–51; and Frank Uekoetter, *The Age of Smoke: Environmental Policy in Germany and the United States, 1880–1970* (Pittsburgh: University of Pittsburgh Press, 2009), 96.

16. For some of the latest publications on this issue, see Robin Bunton and Alan Peterson, eds., *Genetic Governance: Health, Risk, and Ethics in the Biotech Era* (London: Routledge, 2005); and Frank Kempken, Renate Kempken, and Kerstin Stockmeyer, *Gentechnik bei Pflanzen: Chancen und Risiken*, third edition (Berlin: Springer, 2006).

17. Frank Uekoetter, "Know Your Soil: Transitions in Farmers' and Scientists' Knowledge in Germany," in *Soils and Societies: Perspectives from Environmental History,* edited by John McNeill and Verena Winiwarter (Isle of Harris: White Horse Press, 2006), 320–38.

18. Ralf Henneking, *Chemische Industrie und Umwelt: Konflikte um Umweltbelastungen durch die chemische Industrie am Beispiel der schwerchemischen, Farben- und Düngemittelindustrie der Rheinprovinz (ca. 1800–1914)* (Stuttgart: Steiner, 1994).

19. See, for example, Werner Abelshauser, ed., *Die BASF: Eine Unternehmensgeschichte* (Munich: Beck, 2002).

20. Christian Warren, *Brush with Death: A Social History of Lead Poisoning* (Baltimore, Md.: Johns Hopkins University Press, 2000).

21. Uekoetter, *Age of Smoke*.

22. According to Volker Klemm, agricultural science started with Albrecht Daniel Thaer (1752–1828), who drew heavily on his first observations during a prolonged stay in England (see Volker Klemm, "Die Entstehung eigenständiger Landbauwissenschaften in Deutschland [1800–1830]," *Zeitschrift für Agrargeschichte und Agrarsoziologie* 44 [1996]: 162–73). One may quibble about the criteria for his choice, but there can be no general doubt that agricultural expertise grew in both importance and sophistication around 1800.

23. Paul Wagner, *Stickstoffdüngung und Reingewinn* (Berlin: Parey, 1906), 24.

24. Uekoetter, "Know Your Soil."

25. Onno Poppinga, "Änderungen im Kleinen—weiter wie zuvor im Großen: Eine kritische Würdigung der 'Agrarwende' am Beispiel zentraler Themenbereiche," in *Landwirtschaft 2006: Der kritische Agrarbericht*, edited by AgrarBündnis e.V. (Rheda-Wiedenbrück: ABL-Verlag, 2006), 28.

26. David Evans, *A History of Nature Conservation in Britain*, second edition (London: Routledge, 1997), 42; Andreas Knaut, *Zurück zur Natur! Die Wurzeln der Ökologiebewe-

gung (supplement 1 [1993] of *Jahrbuch für Naturschutz und Landschaftspflege*) (Greven: Hilda, 1993), 27; and Alon Tal, *Pollution in a Promised Land: An Environmental History of Israel* (Berkeley: University of California Press, 2002), 167.

27. Edward O. Wilson, *The Future of Life* (New York: Alfred A. Knopf, 2002), xix.

28. Walther Schoenichen, "'Wir wollen einen Naturschutzpark gründen und ...,'" *Naturschutz* 15 (1934): 137.

29. Stefan Körner, "Die Entwicklung des Naturschutzes und der Landschaftsplanung nach dem Zweiten Weltkrieg," in *Natur- und Umweltschutz nach 1945: Konzepte, Konflikte, Kompetenzen*, edited by Franz-Josef Brüggemeier and Jens Ivo Engels (Frankfurt: Campus, 2005), 87–102.

30. Peter Lundgreen, "Akademisierung—Professionalisierung—Verwissenschaftlichung," *Geschichte in Wissenschaft und Unterricht* 53 (2002): 678–87.

31. John Whiteclay Chambers II, *The Tyranny of Change: America in the Progressive Era, 1900–1917* (New York: St. Martin's Press, 1980), v.

32. Daniel T. Rodgers, "In Search of Progressivism," *Reviews in American History* 10, no. 4 (December 1982): 126.

33. Samuel P. Hays, *Conservation and the Gospel of Efficiency* (1959; reprint, Pittsburgh: University of Pittsburgh Press, 1999).

34. On Waring, see Martin V. Melosi, *Garbage in the Cities: Refuse, Reform, and the Environment, 1880–1980* (College Station: Texas Station A & M University Press, 1981), 51–78.

35. Robert J. Morgan, *Governing Soil Conservation: Thirty Years of the New Decentralization* (Baltimore, Md.: Johns Hopkins University Press, 1965); and R. Burnell Held and Marion Clawson, *Soil Conservation in Perspective* (Baltimore, Md.: Johns Hopkins University Press, 1965).

36. Hansjörg Küster, *Geschichte des Waldes: Von der Urzeit bis zur Gegenwart* (Munich: Beck, 1998), 185.

37. Uekoetter, *Age of Smoke*.

38. Richard J. Evans, *Death in Hamburg: Society and Politics in the Cholera Years, 1830–1910* (Oxford: Oxford University Press, 1987).

39. Jürgen Büschenfeld, *Flüsse und Kloaken: Umweltfragen im Zeitalter der Industrialisierung (1870–1918)* (Stuttgart: Kletta-Cotta, 1997), 52–64, 79–87.

40. Marc Reisner, *Cadillac Desert: The American West and Its Disappearing Water* (New York: Penguin, 1987).

41. Albrecht Weisker, "Powered by Emotion? Affektive Aspekte in der westdeutschen Kernenergiegeschichte zwischen Technikvertrauen und Apokalypseangst," in *Natur- und Umweltschutz nach 1945*, edited by Brüggemeier and Engels, 203–21.

42. Ute Frevert, ed., *Vertrauen: Historische Annäherungen* (Göttingen: Vandenhoeck & Ruprecht, 2003).

43. Ernst Schillinger, "Breisach: Der Kampf beginnt," in *Wyhl: Kein Kernkraftwerk in*

Wyhl und auch sonst nirgends. Betroffene Bürger berichten, edited by Bernd Nössler and Margret de Witt (Freiburg: Dreisam-Verlag, 1976), 30.

44. Brian Balogh, *Chain Reaction: Expert Debate and Public Participation in American Commercial Nuclear Power, 1945–1975* (Cambridge: Cambridge University Press, 1991).

45. Peter Weingart, *Die Stunde der Wahrheit? Zum Verhältnis der Wissenschaft zu Politik, Wirtschaft, und Medien in der Wissensgesellschaft* (Weilerswist: Velbrück, 2001), 353.

46. Holger Strohm, *Friedlich in die Katastrophe: Eine Dokumentation über Kernkraftwerke* (Hamburg: Verlag Association, 1978).

47. Egmont R. Koch and Fritz Vahrenholt, *Seveso ist überall: Die tödlichen Risiken der Chemie* (Cologne: Kiepenheuer & Witsch, 1978); and Verband der Chemischen Industrie, ed., *Seveso ist nicht überall* (Frankfurt: Verband der Chemischen Industrie, 1978).

48. R. Allan Freeze, *The Environmental Pendulum: A Quest for the Truth about Toxic Chemicals, Human Health, and Environmental Protection* (Berkeley: University of California Press, 2000), 20.

49. Thomas F. Glick, "Science, Technology, and the Urban Environment: The Great Stink of 1858," in *Historical Ecology: Essays on Environment and Social Change*, edited by Lester J. Bilsky (Port Washington, N.Y.: Kennikat Press, 1980), 122–39.

50. *Die Zeit,* no. 2 (January 5, 2006).

Chapter 10. Desertification

Helpful comments to this chapter from Professor Uriel Safriel are gratefully acknowledged.

1. Lester Brown, "Natural Systems under Stress," in *Plan B 2.0: Rescuing a Planet under Stress and a Civilization in Trouble* (New York: W. W. Norton & Company), 2006, chapter 5.

2. This language is from the United Nations Convention to Combat Desertification (UNCCD).

3. Y. Avni, "Gully Incision as the Main Cause of Ongoing Aridization of the Negev Highlands," in *Proceedings of Israel Geological Society Annual Meeting* (Jerusalem: IGS, 1998), 9.

4. See, generally, Daniel Hillel, *The Natural History of the Bible* (New York: Columbia University Press, 2006).

5. Jared Diamond, *Guns, Germs, and Steel: The Fates of Human Societies* (New York: Norton, 1997), 312.

6. L. Lavauden, "Les Foretsdu Sahara," *Revenue des Eaux et Forets* 7 (1927): 329–41, as quoted in H. E. Dregne, "Land Degradation in the Drylands," *Arid Land Research and Management* 16 (2002): 132.

7. A. Aubreville, "Climats, forets et désertification de l'Afrique tropicale," *Société d'Editions Géographiques: Maritimes ent Coloniales* (1949), as quoted in Michael Mortimore, *Roots in the African Dust: Sustaining the Drylands* (London: Oxford University Press, 1998), 341.

8. Millennium Ecosystem Assessment, *Ecosystems and Human Well-Being: Desertification Synthesis* (Washington, D.C.: World Resources Institute, 2005).

9. Quoted in Marc Reisner, *Cadillac Desert: The American West and Its Disappearing Water* (New York: Penguin, 1993), 149–51.

10. Public Law No. 74-46 Stat. 163 (1935), currently 16 U.S.C. sec. 590. For a history of U.S. soil erosion legislation, see John Braden and Donald Uchtmann, "Soil Conservation Programs amidst Faltering Environmental Commitments and the 'New Federalism,'" *Boston Environmental Affairs Law Review* 10, no. 639 (1982): 639–96.

11. See, generally, Alon Tal and Jessica Cohen, "Adding 'Top Down' to 'Bottom Up': A New Role for Environmental Legislation in Combating Desertification," *Harvard Journal of Environmental Law* 31, no. 1 (2007): 163–219.

12. Joel Cohen, *How Many People Can the Earth Support?* (New York: W. W. Norton, 1975), 79.

13. See, for example, Mortimore, *Roots in the African Dust*.

14. William Ellis, "Africa's Sahel: The Stricken Land," *National Geographic* 172, no. 2 (August 1987): 140–79.

15. Alex Shoumatoff, "Forecast: Dry and Brutal," *One Earth* (Winter 2006): 28.

16. William C. Burns, "The International Convention to Combat Desertification: Drawing a Line in the Sand?" *Michigan Journal of International Law* 16 (1995): 831.

17. Simcha Blass, *Water in Strife and Action* (Givataim, Israel: Masada, 1973), 330.

18. Daniel Hillel, "Irrigated Agriculture," in *The Negev: Land, Water, and Life in a Desert Environment* (New York: Praeger, 1982), 221.

19. Blass, *Water in Strife and Action*, 351.

20. Hillel, "Irrigated Agriculture," 221.

21. Alon Tal, "Seeking Sustainability: Israel's Evolving Water Management Strategy," *Science* 313 (2006): 1081–84.

22. Fred Pearce, *When Rivers Run Dry: Water, the Defining Crisis of the Twentieth Century* (Boston: Beacon, 2006), 202–4.

23. Marq de Villiers, *Water* (Toronto: Stoddart, 2002), 130–35.

24. Pearce, *When Rivers Run Dry*, 201.

25. Ibid.

26. Ibid.

27. World Commission on Environment and Development, *Our Common Future* (London: Oxford, 1987).

28. Report of the Conference of the Parties on Its Seventh Session, Held in Nairobi, from October 17 to 28, 2005, available online at http://www.unccd.int, document reference: Iccd/Cop(7)/16/ 25 November 2005.

29. *UNCCD News*, July–August 2009 edition, available online at http://newsbox.unccd.int/1.1/.

30. Millennium Ecosystem Assessment, *Ecosystems and Human Well-Being*, 7.

31. Lester Brown, "China Losing War with Advancing Deserts," *EcoEconomy Updates*, available online at http://www.earth-policy.org/Updates/Update26.htm (accessed on August 5, 2003).

32. Ibid., at section 2.3, p. 5.

33. China Ministry of Forestry, *A Rising Green Great Wall* (Beijing: Dadi Publishing House, 1988).

34. The Order of the President of the People's Republic of China (No. 55), "The Law of Desertification Prevention and Control of the People's Republic of China, Aug. 31, 2001" (hereafter referred to as "China Desertification Law"), available online at http://www.adb.org/Projects/PRC_GEF_Partnership/Desertification.pdf (accessed on May 19, 2010).

35. Ibid., art. 38.

36. Ibid., art. 43.

37. Ibid., art. 34.

38. See the Web site of the International Year of Deserts and Desertification, at http://www.iydd.org.

39. Ibid.

40. Victor Louro and Ana Maria Almeida, "Portugal Report to Regional Meeting of Annexes IV and V," presentation delivered at the meeting of UNCCD Annex I countries, Bonn, Germany, July 2006.

List of Contributors

JENS IVO ENGELS is a professor of modern and contemporary history at Technische Universität in Darmstadt, Germany. His books include *Northern Europe: An Environmental History* (2005, with Tamara L. Whited, Richard C. Hoffmann, Hilde Ibsen, and Wybren Verstegen) and *Naturpolitik in der Bundesrepublik: Ideenwelt und politische Verhaltensstile in Naturschutz und Umweltbewegung, 1950–1980* (2006).

DEBORAH FITZGERALD is a professor in the History of Technology Program in Science, Technology, and Society at MIT. Her books include *Every Farm a Factory: The Industrial Ideal in American Agriculture* (2003) and *The Business of Breeding: Hybrid Corn in Illinois, 1890–1940* (1990).

BERND-STEFAN GREWE is the director of the research group Dynamics of Transnational Agency at the Cluster of Excellence at the University of Konstanz in Germany. His research interests include environmental history (*Der versperrte Wald*, 2004), global and international history (*Project: The Global History of Gold in the Twentieth Century*), and material culture (*Luxury in Global Perspective: Objects and Practices*, 2010).

THOMAS LEKAN is an associate professor of history at the University of South Carolina in Columbia and currently a residential fellow at the National Humanities Center in Research Triangle Park, North Carolina. His books include *Imagining the Nation in Nature: Landscape Preservation and German National Identity, 1885–1945* (2004) and, with coeditor Thomas Zeller, *Germany's Nature: Cultural Landscapes and Environmental History* (2005).

J. R. McNEILL teaches world history, environmental history, and international history at Georgetown University. His books include *Something New under the Sun: An Environmental History of the Twentieth-century World* (2000) and *Mosquito Empires: Ecology and War in the Greater Caribbean, 1640–1914* (2010).

CHRISTIAN PFISTER is an emeritus professor of economic, social, and environmental history at the University of Bern. Besides publishing several books in German, he has written articles on population, climate, disaster, and environmental history, such as "Social Vulnerability to Climate in the Little Ice Age: An Example from Central Europe in the Early 1770s" (2006) and "Climatic Extremes, Recurrent Crises, and Witch Hunts: Strategies of European Societies in Coping with Exogenous Shocks in the Late Sixteenth and Early Seventeenth Centuries" (2007). For the list of his publications and downloads, see http://www.wsu.hist.unibe.ch.

ALON TAL is a professor at the Jacob Blaustein Institute for Desert Research of Ben-Gurion University of the Negev in Israel. His books include *Pollution in a Promised Land: An Environmental History of Israel* (2002) and *Speaking of Earth: Environmental Speeches That Moved the World* (2006).

JOEL A. TARR is Richard S. Caliguiri University Professor of History and Policy at Carnegie Mellon University in Pittsburgh. He has published *The Search for the Ultimate Sink: Urban Pollution in Historical Perspective* (1996) and, with Clay McShane, *The Horse in the City: Living Machines in the Nineteenth Century* (2007).

FRANK UEKOETTER is a Dilthey Fellow with the Research Institute of the Deutsches Museum in Munich and deputy director of the Rachel Carson Center, also in Munich. His books include *The Green and the Brown: A History of Conservation in Nazi Germany* (2006) and *The Age of Smoke: Environmental Policy in Germany and the United States, 1880–1970* (2009).

Index

agriculture, 13, 16–17, 19, 22–23, 29–43, 109–10, 136, 138, 151; productivity of, 30, 32, 109–11, 136, 146–47
Allmen, Peter von, 103
Andersen, Arne, 94
animal husbandry, 17, 26, 30, 35, 138
Anthropocene, 10, 92
Aral Sea, 65, 153–54
Archer Daniels Midland, 25–27
Argentina, 26
Australia, 14–16, 26–27, 59, 67, 97, 150

Bacon, Francis, 132–33, 141
Bagwell, Philip, 113
Bairoch, Paul, 109
Beck, Ulrich, 120–21
Bennett, Hugh, 133
Binswanger, Hans-Christoph, 94
Black Death, 24
Blass, Simcha, 152
Bonne, Georg, 142
Borchardt, Knut, 99–100
Borscheid, Peter, 113
Braudel, Fernand, 11
Brazil, 25
Bretton Woods, 99–100
Brown, Lester, 158
Brunhes, Jean, 134
Bureau of Reclamation, 34, 62, 140, 142–43

California, 27, 35, 38–40, 85, 135
Canada, 26, 149
carbon dioxide, 21, 90–91, 103
Caribbean, the, 7, 25–26, 29–31
Carson, Rachel, 135
Chakrabarty, Dipesh, 117
Charcoal, 20–26, 45, 48–49, 51
China, 20, 22–23, 25, 67, 69, 146, 158–60

cholera, 76, 142
cities, 13, 16, 20–21, 23, 28, 72–89; agriculture, 21; climate change, 21; effects on environment, 72, 77; migration to, 20; mortality rates, 20, 79–80; natural disasters, 74, 89; pollution, 76, 81, 83–84; population of, 73, 78, 81, 84, 86–87
Civilian Conservation Corps, 37
Clean Air Acts, 85
Clean Water Act (1972), 85
climate change, 2, 7, 11, 15–16, 21, 58, 106, 117, 128, 145
coal, 26, 49, 53, 76, 83, 103–4, 134, 137, 140
Columbian Exchange, 6, 13, 23–25, 28
COMECON, 104
Crafts, Nick, 96, 99
Crosby, Alfred, 6, 23
Crutzen, Paul, 10, 92

dams, 62–66, 69, 81
De Grazia, Victoria, 106
deforestation, 25, 27
DeGolyer, Lee Everett, 101
Denmark, 110, 145
desertification, 4, 11, 146–60. *See also* Dust Bowl
Diamond, Jared, 147
Dienel, Hans Luidger, 133
disease, 15–19, 22–24, 26, 56, 76, 79, 142
domestication, 15–20, 22–24, 28
Dongtan, 69–70
Dregne, Harold, 157
Duchess of Portland, Winifred, 138
Dust Bowl, 9, 36, 43, 55, 62, 140, 149

Egypt, 26, 150
Eisenhower, Dwight, 103
Elvin, Mark, 22

England, 20, 54, 59, 138, 149,
environmental movement, 5, 11, 84, 109, 111, 119, 145; and history, 119–20; impact of, 128; features of, 124; non-state actors, 127, 138; politicization, 125–27; science, 125, 128; as turning point, 119–20, 123
Environmental Protection Agency, 85
Ethiopia, 150–51
evolution, 14–15, 18

famine, 136, 148, 151
fertilizer, 27, 30, 38, 75, 109–10, 112, 136, 152
fire, 13–14, 28
Floh, Hermann, 91
forest, 7, 22, 44–54, 128; change over time, 44–45, 51–52; commercialization, 49, 140; economics of, 48–49
forestry, 7, 45, 47, 58, 67, 128, 140; effect on forests, 52, 59; origins of, 47–48, 50, 59; in twentieth century, 51;
Forestry Management Act (UK, 1919), 61
fossil fuels, 7, 13, 25–28, 58, 83, 97, 111–12, 114, 134; consequences of, 117; effects on cities, 26; and labor, 26; price, 92, 100, 105. *See also* coal; oil
Fourastié, Jean, 96
France, 7, 38, 50, 53, 59, 68, 96, 103, 107, 110, 112, 116, 127
Friedrich, Caspar David, 59

General Agreement on Tariffs and Trade, 99
genetics, 37, 40–41, 130, 135
Giddens, Anthony, 120
Gingrich, Simone, 104
Glacken, Clarence, 60
globalization, 7, 11, 25, 54, 56, 116
greenhouse effect. *See* climate change

Haagen-Smit, Arie, 85
Hambach Castle, 44–46
Hatch Act (1887), 33, 40
Haustein, Sabine, 112
Hays, Samuel, 84, 140
Heymann, Mathias, 129
Hicks, John, 105
higher education, 32–33, 99
Hobsbawm, Eric, 99, 111
Homeland Protection League, 65–66
Homestead Act (1862), 33

horticulture, 16, 32
hunter-gatherer, 19–20, 22

Iliffe, John, 22
India, 22–23, 26, 56, 59, 64, 67, 69, 140, 150
industrialization, 3, 5, 8, 25–26, 38, 50, 72, 92, 95, 117, 123, 141
Interstate Highway Act (1956), 86
Iran, 101
Iraq, 27, 101
irrigation, 16, 21, 33, 38, 64, 152–53
Israel, 138, 147, 152–53, 160
Italy, 113, 144

Jaffe, Adam, 105
Japan, 25, 96–97, 100
Josephson, Paul, 62–63

Kenya, 150, 152
Koch, Egmont, 144
König, Wolfgang, 94
Krausmann, Fridolin, 97, 104, 110
Kupper, Patrick, 94, 124
Kuskova, Petra, 104

lactose tolerance, 18
language, 13–14, 28
Liebig, Justus von, 32
livestock. *See* animal husbandry
Lomborg, Bjørn, 122, 145
Lorenz, Konrad, 126
Lundgreen, Peter, 139

Maier, Charles, 8, 56, 58, 61
malaria, 18–19
Mannion, Antoinette, 112
Marsh, George Perkins, 67
Marshall Plan, 99, 110
Massarat, Mohssen, 99
McCormick, Cyrus, 36, 41
mechanization, 35, 36, 38–41, 109, 111
Mendel, Gregor, 32, 37
Merchant, Carolyn, 1, 58, 132–33
Merki, Christoph Maria, 95, 113
methane, 21
Mexico, 25, 27, 29
migration, 13–14, 43; effect on flora and fauna, 15; out of Africa, 14–15, 28; to cities, 20, 50; westward (in U.S.), 33, 42

INDEX

mining, 25–27
Mintz, Sidney, 31
Mom, Gijs, 113–14
monoculture, 29, 34, 36–39; and forests, 51, 59
Monsanto, 41
Morrill Land Grant Act, 33
Muir, John, 65

Nash, Roderick, 60
Native Americans, 23–24
Neolithic Revolution, 2–3, 16–19
Netherlands, 49, 54, 68, 108, 153, 158
New Deal, the, 9, 34, 63, 82, 140
Newell, Richard, 105
Nigeria, 150, 154
night soil, 21
"1950s syndrome," 4, 10, 54, 90–118, 122

oil, 27, 99–104, 117, 140
OPEC, 100
Osborn, Fairfield, 126
overpopulation, 126, 148. *See also* population

pastoralists, 20, 22
Peasant War (1525), 45–46, 53
pesticides, 27, 109
plant breeding, 17, 37, 41
Poland, 21
Polanyi, Karl, 92
political ecology, 126
Pollard, Sydney, 5
pollution, 25–26, 54, 58, 72, 87, 121–24, 129, 135–36, 142; air, 27, 82–83, 122, 137–38, 140; water, 27, 38, 76, 123
Popp, David, 104
population, 19, 49, 81–82, 87, 97–98, 111, 150–51, 153. *See also* overpopulation
Priebe, Hermann, 110

radioactivity, 68
Radkau, Joachim, 95
railroad, 26, 33, 35, 49, 54, 62, 73, 104, 113; cargo transportation, 115; cities and, 75; popularity of, 117
Rees, William, 107
Resource Conservation and Recovery Act (1976), 85
Revelle, Roger, 90
Rhine River, the, 44, 59–61, 65–66, 69; pollution, 68–69, 123; wildlife of, 69, 123

Rodgers, Daniel, 139
Romanticism, 60–61, 67–68, 70, 129
Roosevelt, Franklin, 150
Ruddiman, William, 21
Rudorff, Ernst, 138
Rural Resettlement Association, 62

Safe Drinking Water Act (1974), 85
Saudi Arabia, 101, 117, 147
Schildt, Axel, 94
Scott, James, 32, 56–58, 61–63, 65, 70, 134
sewage, 75, 79–82, 84, 87, 123, 135
Sieferle, Rolf Peter, 2–3, 95, 109
Siegenthaler, Hansjörg, 94, 100
Sierra Club, 65
silk road, 23
slash-and-burn agriculture, 151
slavery, 25–26, 31, 59
Smith-Lever Act (1914), 33
Söderbaum, Peter, 94
soil chemistry, 32, 136
Soil Conservation Service, 62, 133, 140, 150
Soil Erosion Act (1935), 150
solar energy, 106
Sombart, Werner, 48, 66
Soviet Union, the, 56, 62–65, 69, 97–98, 104, 153–54
Spain, 23, 116, 150
standard of living, 107
state, the, 8, 11, 23, 32–33, 41, 55, 85, 123; forests and, 47–48, 51, 59; impact on environment, 56, 58, 64–65, 70–71, 141; power of, 61, 126
Stavins, Robert, 105
Steffen, Will, 92–93, 95
Steinbeck, John, 149
Stern, Nicholas, 117
Strohm, Holger, 144
Suess, Hans, 90
sugar, 25–26, 29, 31
Superfund legislation, 85
Sutter, Paul, 67

Texas, 27, 87
Thirsk, Joan, 42
Thoreau, Henry David, 138
tourism, 60–61, 116
transportation, 26, 35, 39, 73, 113, 115, 129; air, 116–17; canals, 73; car, 82–83, 86, 88, 106, 111, 113–15, 118; local, 73, 80, 88; steamship, 26, 104. *See also* railroad

Traube, Klaus, 144
Tulla, Johann Gottfried, 59–60

Vahrenholt, Fritz, 144
Venezuela, 27
Vogt, William, 126

Wackernagel, Mathis, 107
Wagner, Paul, 138
war, 38–39, 61, 127
water systems, 74–75, 78, 80–82

Weingart, Peter, 144
White, Lynn, 2
Wilson, Edward, 138
Wittfogel, Karl, 63
wood, 20, 47–50, 52–53, 76
World Bank, 41
Worster, Donald, 55, 63

yellow fever, 17, 76
Yellowstone National Park, 60
Yoffee, Avraham, 138